ROUTLEDGE LIBRARY EDITIONS: URBAN PLANNING

Volume 22

PLANNING IN EUROPE

PLANNING IN EUROPE
Urban and Regional Planning in the EEC

Edited by
R. H. WILLIAMS

LONDON AND NEW YORK

First published in 1984 by George Allen & Unwin (Publishers) Ltd

This edition first published in 2018
by Routledge
2 Park Square, Milton Park, Abingdon, Oxon OX14 4RN

and by Routledge
711 Third Avenue, New York, NY 10017

Routledge is an imprint of the Taylor & Francis Group, an informa business

© 1984 R. H. Williams

All rights reserved. No part of this book may be reprinted or reproduced or utilised in any form or by any electronic, mechanical, or other means, now known or hereafter invented, including photocopying and recording, or in any information storage or retrieval system, without permission in writing from the publishers.

Trademark notice: Product or corporate names may be trademarks or registered trademarks, and are used only for identification and explanation without intent to infringe.

British Library Cataloguing in Publication Data
A catalogue record for this book is available from the British Library

ISBN: 978-1-138-49611-8 (Set)
ISBN: 978-1-351-02214-9 (Set) (ebk)
ISBN: 978-1-138-48565-5 (Volume 22) (hbk)
ISBN: 978-1-138-48569-3 (Volume 22) (pbk)
ISBN: 978-1-351-04840-8 (Volume 22) (ebk)

Publisher's Note
The publisher has gone to great lengths to ensure the quality of this reprint but points out that some imperfections in the original copies may be apparent.

Disclaimer
The publisher has made every effort to trace copyright holders and would welcome correspondence from those they have been unable to trace.

Planning in Europe
Urban and Regional Planning in the EEC

Edited by
R. H. WILLIAMS
Lecturer in Town and Country Planning,
University of Newcastle upon Tyne

London
GEORGE ALLEN & UNWIN
Boston   Sydney

© R. H. Williams, 1984.
This book is copyright under the Berne Convention. No reproduction without permission. All rights reserved.

**George Allen & Unwin (Publishers) Ltd,
40 Museum Street, London WC1A 1LU, UK**

George Allen & Unwin (Publishers) Ltd,
Park Lane, Hemel Hempstead, Herts HP2 4TE, UK

Allen & Unwin, Inc.,
9 Winchester Terrace, Winchester, Mass. 01890, USA

George Allen & Unwin Australia Pty Ltd,
8 Napier Street, North Sydney, NSW 2060, Australia

First published in 1984.

British Library Cataloguing in Publication Data

Planning in Europe: urban and regional
planning in the EEC.—(Urban and regional
studies; no. 11)
1. Regional planning—European Economic
Community countries
I. Williams, R.H.   II. Series
361.6′1′094   HT391
ISBN 0-04-711012-0

Library of Congress Cataloging in Publication Data

Planning in Europe.
(Urban and regional studies; no. 11)
1. City planning—European Economic Community
countries—Addresses, essays, lectures. 2. Regional
planning—European Economic Community countries—Addresses,
essays, lectures. I. Williams, R. H. (Richard Hamilton),
1945-   . II. Series: Urban and regional studies
(George Allen & Unwin); no. 11.
HT169.E85P58 1984   307′.12   83-22374-
ISBN 0-04-711012-0

Set in 10 on 11 point Times by V & M Graphics Ltd, Aylesbury, Bucks
and printed in Great Britain
by Billing and Sons Ltd, London and Worcester

Contents

Notes on the Contributors		page	ix
Acknowledgements			x
Preface			xi
1	Introduction *by R. H. Williams*		1
2	The Federal Republic of Germany *by K. R. Kunzmann*		8
3	Italy *by S. Bardazzi*		26
4	France *by P. M. B. Chapuy*		37
5	The Netherlands *by H. van der Cammen*		49
6	Belgium *by M. Anselin*		63
7	Luxembourg *by N. von Kunitzki*		73
8	The United Kingdom *by R. H. Williams*		86
9	Ireland *by K. I. Nowlan*		103
10	Denmark *by O. Kerndal-Hansen*		114
11	Greece *by A.-Ph. Lagopoulos*		128
12	The European Communities *by R. H. Williams*		144
13	Some Links and Comparisons *by R. H. Williams*		159
Bibliography			172
Index			185

Notes on the Contributors

PROFESSOR DR M. ANSELIN is head of the Seminarie voor Survey en Ruimtelijke Planning in the State University of Gent.

PROFESSOR S. BARDAZZI is head of the Istituto Riccerca Territoriale e Urbana in the University of Florence.

DR HANS VAN DER CAMMEN is a lecturer in the Department of Planologie en Demografie in the University of Amsterdam.

PIERRE CHAPUY is a consultant with the Groupe d'Etudes Ressources Planification Aménagement in Paris.

OLE KERNDAL-HANSEN is a member of the Institut for Center-Planlaegning, Copenhagen.

DR N. VON KUNITZKI is president of the Centre for Juridical Studies and Comparative Law, Institut Universitaire International, Luxembourg.

PROFESSOR DR KLAUS R. KUNZMANN is head of the Institut für Raumplanung, University of Dortmund.

PROFESSOR DR A.-PH. LAGOPOULOS is Professor of Town Planning in the University of Thessaloniki.

PROFESSOR K. I. NOWLAN is Professor of Regional and Urban Planning at University College, Dublin.

Acknowledgements

The speech quoted in Chapter 9 is reproduced from *Parliamentary Debates* of 31 July 1963, with the permission of the Controller, Stationery Office, Dublin, Ireland.

Professor Lagopoulos would like to thank all who provided him with indispensable data for the completion of Chapter 11, and especially Mr G. Kafkalas, Mr N. Katochianos, Dr I. Michael and Mr A. G. Romanos for their helpful comments on the manuscript.

Preface

In compiling this book the aim has been to assemble within one volume a review of the systems of town and country planning that operate within the member-states of the European Economic Community, and to link this with an outline of the ways in which the community's institutions and policies relate to the activity of planning. The popularity of overseas study visits for students of town and country planning, geography and other disciplines, and contributions to the academic and professional press, reflect a widespread desire to learn by studying planning practice in other countries. It is hoped that this book will be of value to students, practitioners and officials wanting to gain an understanding of planning elsewhere in Europe.

As a matter of deliberate editorial policy each of the chapters of this book that relates to a specific country is written by an expert contributor native to that country. The contributors were invited to concentrate on the scope and style of planning in their country, rather than provide a detailed account of planning legislation as such; and it was felt that a critical impression of the flavour of planning in each country would be better conveyed by a writer from that country, rather than from an outside observer. The approaches adopted by the contributors varies not only as a result of their own judgement of what aspects of their subject deserve to be brought most prominently to the readers' attention, but also because the extent to which a country's planning system is already internationally known varies considerably. Where suitable further reading is available, this is mentioned.

I should like to record here my thanks to all the contributors for their co-operation and interest in this project, not to mention the remarkable command of English shown by most of them.

Inspiration for this book arose out of my work on the International Affairs Board of the Royal Town Planning Institute, on which I have served since 1977. I owe a particular debt of gratitude to Professor Gordon Cherry of the Centre for Urban and Regional Studies, University of Birmingham. As president of the Royal Town Planning Institute in 1978-9 he initiated contacts with professional associations of planners in the other EEC countries, and during his presidential year found time to give me direct encouragement to develop ideas in a paper prepared by me in support of this initiative into the form of a book.

I must also acknowledge my debt to the postgraduate and undergraduate students attending my lecture course on comparative planning in the EEC, in the University of Newcastle upon Tyne, whose enthusiasm and response to the course has convinced me of the value of a book such as this, and in numerous ways influenced my editorial judgement.

I have also benefited greatly from discussions with many academics in Britain and Europe, who in various ways have shaped my ideas. It would be impossible to list everyone but I must make particular mention of the assistance I have received from Dr Uwe Wullkopf and his colleagues at the Institut Wohnen and Umwelt in Darmstadt, who have on two occasions accommodated me when I have made study visits to Germany.

I must record here my gratitude to Anne Hudson, who typed the text with great skill and efficiency, and to Mia Wilkins and Paolo Scattoni for assistance with translation.

Finally, I must pay tribute to the support and encouragement of my wife, and toleration of my children, during the preparation of this book. I am fortunate that they all share in various ways my own fascination for Europe.

URBAN AND REGIONAL STUDIES NO. 11

Planning in Europe

Chapter 1

Introduction

R. H. Williams

In recent years there has been a growing interest in studying the planning systems of other countries, and of comparative studies particularly related to Europe. Perhaps, in the case of Britain, this represents a belated response to entry into the European Economic Community (EEC) in 1973. By presenting accounts of the planning systems of the member-states it is hoped that a useful starting-point will be provided on which to have such studies. The bulk of this book consists of these accounts. In this chapter the context is set by discussing the rationale underlying the editorial policy pursued here, and the purposes which it is hoped this book will serve.

The order of the chapters devoted to individual countries is inevitably somewhat arbitrary. The original members of the EEC come first, roughly in order of the size of their population (Chapters 2–7). These are followed by the three who joined in 1973, the United Kingdom, Ireland and Denmark (Chapters 8–10), and then comes the chapter on the most recent newcomer to the Community, Greece, who joined on 1 January 1981 (Chapter 11). The two remaining chapters of the book are devoted to the EEC itself and an attempt to draw together some conclusions from the preceding material.

Chapter 12 is an account of the institutions and policies of the European communities which are related to the practice of town and country planning, directly or indirectly. The interpretation of this relationship is, perhaps inevitably, based on a British perception of the role of a local planning authority. Although the EEC exists primarily to achieve economic and political objectives, its initiatives and activities have implications for the practice of planning, and will do so increasingly during the 1980s, as the discussion in Chapter 12 indicates.

It is in the nature of a book such as this, written by a number of authors and spanning a wide range of subject-matter, that many topics are described and ideas introduced which do not lend themselves easily to synthesis and summary in the form of conclusions. Nevertheless, the final chapter does attempt to draw the threads together, to some degree, in order to identify some common elements, important distinctions, or points of comparison which may have escaped attention in the accounts of individual countries (Chapter 13). It is not intended or claimed that this will form a thorough comparative analysis of all the planning systems treated here. Rather, it aims to provide a basis on which further work of a comparative nature might be developed.

WHY THE EEC?

The decision to confine the choice of countries included here to those that are members of the EEC was taken not simply because this represented a convenient limitation, although of course it is necessary to have some criteria of selection for purely practical reasons. More importantly, it was decided to concentrate attention on the members of the EEC, because they are in supra-national association and have certain characteristics in common. All have democratic forms of government and a mixed economy, and all have adopted legislation providing for some form of regulation of development and land-use planning. Furthermore, in joining the three European Communities, that is, the European Economic Community (EEC), the European Coal and Steel Community (ECSC) and Euratom, all the member-states agree to adopt certain common policies and rules in addition to their own national policies and legislation. In a number of respects these common policies and rules have a bearing on the practice of planning and the concerns of local planning authorities.

Many of the Commission's policies are directed towards establishing conditions of fair competition; stimulating trade within the Community; overcoming disparities in economic performance between different regions of the Community; and improving the transport infrastructure, especially where this might assist in the movement of goods between member-states. The Commission is also anxious to protect the environment – reflecting the growing popular and political pressure in this direction in Europe – and is making progress towards the adoption of measures designed to protect environmental quality and ensuring that no major variations exist in the degree of control over development for reasons of environmental protection. These measures will relate directly to the widely varied systems of planning that have been established by member-states over the years.

A further way in which the EEC may have an influence on planning is by means of measures which affect the planners themselves. Freedom for people to seek employment and practice their profession is a principle embodied in the Treaty of Rome. There are a number of planners working outside their own country, or who have clients from other countries, but the numbers involved are not great. In a subject such as planning where so much of the style of professional practice is determined by political and cultural factors, it is not sufficient to possess a basic body of purely technical knowledge, or knowledge of the appropriate language, in order to work elsewhere.

Whether or not any Community-wide interest group representing the planning profession as a whole becomes established in addition to the existing committee looking after the interests of planners in private practice, planners throughout the EEC will have to respond to Community initiatives when initial consultations are taking place, and implement whatever measures are adopted. Since any such measures are likely, at least when first proposed, to fit into one country's planning and legal systems more neatly than another's, it follows that the reaction of planners and others affected is likely to be influenced by the degree to which a proposal is complementary to their own system of planning. An understanding of the background influences which operate in European planning systems, and the role of the Commission, are valuable prerequisites for constructive criticism of any Commission proposal, and increase the likelihood that any proposals eventually adopted will be in acceptable and useful form.

EVOLUTION OF PLANNING SYSTEMS AND COMPARATIVE STUDY

The countries of Western Europe have all, in the postwar era, experienced a high degree of urbanisation. Some, like the UK, have been highly urbanised for a relatively long time, while others such as Italy and Greece have experienced this phenomenon much more recently. Nevertheless, it is generally true that urbanisation and industrialisation is typical of the EEC. Associated with this have been increases in personal wealth and car ownership rates, higher expectations regarding housing quality and a movement of population away from employment in agriculture towards industrial or service occupations. It is not the task of this book to describe in detail the trends in urban development or regional differentiation that have taken place in Europe in recent years, but rather to note that these processes have all had the effect of increasing the competing pressures on the land. Pressures to house the growing urban population, often at decreasing densities as more people aspire to individual dwellings; pressures to accommodate industry or offices, often for reasons of scale, possible hazard, or environmental effects, on sites removed from residential areas; and pressures to develop transport networks to link these separated land uses, allowing the exploitation of private motor transport as well as the advantages of mass public transport, and providing for national and international movement of people and goods, have all followed from the urbanisation and industrialisation of Europe.

These pressures on land have been so considerable in the postwar period that all the countries considered here have deemed it necessary to establish some effective procedure to channel these pressures and resolve conflicts between competing land uses. Each country has, therefore, enacted legislation at some time establishing the principle that public authorities should be empowered to monitor and control development and prepared plans which indicate the forms in which acceptable development might take place. Legislation conferring powers of this nature has first appeared at different times in different countries, spanning the late nineteenth to mid-twentieth centuries. These differences reflect differing political attitudes to the acceptability of such powers, which may be regarded as infringing individual rights to enjoy private property, and varying perceptions of the value of planning. They do not necessarily correlate with differing rates of urbanisation or pressure on the land.

The fascination of taking an international view of planning – particularly in the context of the developed nations of Western Europe – lies in the great variety to be found within the town and country planning systems that have been established in response to the common experience of urbanisation, industrialisation and associated pressures on land. This variety is apparent both in the systems of planning and their associated legal procedures, and in the policies and priorities that are pursued. Much can be learned from studying these differences, and the consequent patterns of land use and development in the countries concerned. It is neither easy nor sensible to attempt to say which system of planning as a whole is better than any other, and this is certainly not the intention here. Although, in general terms, the objectives of town and country planning in different countries may be similar, they are not by any means identical, and whatever is appropriate to one country is not necessarily appropriate to another.

Nevertheless, many planning problems are shared by more than one country, and the study of different techniques, procedures, or policies that are brought to bear on particular problems, and their rationale and outcome, is a natural subject of interest to any planner displaying academic curiosity. There is, of course, much more to the comparative study of planning systems than this. The most basic and obvious justification for studying other planning systems is the hope of improving practice in one's own country. This might be achieved by learning lessons from the experience of adopting specific measures in other countries, or by identifying ideas, techniques, or procedures that might usefully be adapted and applied to one's own country. Such transfer is a highly complex business, however, as is apparent when one considers the implications of seeking such a transfer.

There is often a degree of public dissatisfaction with the outcome of planning policies, which in turn stimulates planners to seek improvements. At the same time, there is the problem that one cannot run experiments repeatedly, as in the case of a laboratory-based science, in order to test whether the desired effect is achieved by a new policy before applying it to the community at large. At best, planners may have only one or two opportunities to put a new idea into practice. Evaluation of an innovation is frequently based on limited experience before it is either generally adopted or abandoned. The study of experience elsewhere, observing how similar problems are tackled with different policies, is a valuable method of broadening the base of experience and evaluating ideas that might form the basis of a planning innovation in one's own country. Therefore, many cross-national comparative planning studies have as their objective – explicitly or implicitly – the exploration of the possibility of transfer of policy. A good example of a study with this explicit objective is the Trinational Inner Cities Project, conducted jointly by the University of Reading and research institutes in West Berlin and Washington, DC, in 1978–9.[1]

The practice of town and country planning in any country is conditioned to a great extent by the cultural, governmental, legal and constitutional circumstances of that country, and to disentangle these in order to transfer ideas from one country to another beyond a superficial level is a highly complex business. It is possible to learn something of planning in other countries by observation, or sightseeing, but appreciation of what is observed is likely to remain fairly superficial unless the observer has an understanding of the processes by which particular solutions are achieved, the agencies involved and the political influences on them. Without this background understanding, successful transfer of ideas in many aspects of planning is not likely to be possible.

For example, simple observation of city-centre pedestrianisation in German cities may be, and has been, sufficient stimulus to assist British planners to develop similar ideas in Britain, but it is not possible to form any useful conclusions about potential transfer value of German ideas on gradual renewal of older housing areas without a detailed understanding of the agencies and institutions involved. In order to study this and many other issues for the purpose of comparative analysis and consideration of the possibility of transfer it is necessary to keep the complexities within manageable limits. One way of achieving this is to limit such detailed studies to bilateral comparisons.

Some indication of the complexity of comparative assessment of transfer value of this and many other similar examples will be apparent from reading the various contributions in this book. Although comparative analysis of the depth necessary is clearly beyond the scope of this book, it is hoped that lines of inquiry which might prove fruitful will become apparent to the reader.

While the potential to learn from the experience of others remains the most tangible objective of comparative studies, experience also suggests that one of the greatest benefits to be obtained from pursuing such studies lies not so much in the knowledge gained of other systems as in the opportunity it presents for critical appraisal of one's own system. It is very easy to become so familiar with the approach to certain problems or the procedures adopted in one's own country that it is difficult to conceive of alternative approaches. Study of other systems tends to raise thought-provoking questions about aspects of planning which might otherwise be taken for granted. For instance, in Britain decisions regarding the control of development are taken on the basis of a certain degree of discretion and are not predetermined by allocations that may appear on a development plan. This degree of discretion does not always exist elsewhere. Does this offer Britain valuable flexibility allowing local authorities to respond to priorities prevailing at the time of decision, or does it introduce a degree of uncertainty that is inhibiting to developers? In many countries the mandatory level of development plan is some form of local plan, whereas in Britain it is the structure plan. Why should this be so? These and many other questions could be posed, once one moves outside the confines of one national system of planning.

EDITORIAL POLICY

The countries we consider display not only a very wide variety of planning systems, but also a wide variety of perceptions of the nature and scope of planning, and the range of issues to which planning can legitimately address itself. As a matter of deliberate policy it was decided to give expression to this variety, and to attempt to convey to the reader a sense of the style, scope and practice of planning in each country, and not simply define the legal provisions. Ultimately, of course, the only way in which one can appreciate another planning system in this way is by informal contact with the practitioners in each country. However, this is often not possible. In order to provide a convenient, if limited, substitute for direct contact contributions were invited from experts native to the countries concerned. All are familiar not only with the system of planning in their own country, but also the problems involved in presenting an account to an audience from other countries.

Each contributor was given a common brief, but was allowed considerable latitude to make his own selection of the main themes worthy of emphasis. The main general requirement was to discuss the scope and style of planning practice, the principal issues facing planning authorities, and the procedures and policies adopted in response. In addition, they were invited to discuss the agencies responsible for planning, the technical expertise normally called upon by them, the role of professional planners and politicians within the planning process, and the legal and constitutional basis of planning. There is a constant problem in any international review of a cultural and political phenomenon such as planning of how far to extend the discussion of political and administrative organisation. Contributors embarked on this to the extent that they considered necessary in order to enable the reader to understand the system of planning being described. Likewise, no rigid definition of town and country planning was imposed on the contributors, and the contents of each chapter therefore reflects the author's own interpretation of what is understood by the term town and country planning in his own country. One unavoidable constraint in this respect must be acknowledged, however. The English terminology does not correspond exactly with the technical terms used in other countries to describe particular aspects of planning. For instance, the German word *Raumplanung* or the French phrase *aménagement de territoire* do not correspond precisely with any English equivalent. Consequently, some limitation is imposed by the use of the medium of English.

A full selection of references and suggestions for further reading are grouped together at the end in a classified Bibliography. There are great variations in the extent to which the planning systems referred to are discussed elsewhere in English, or indeed in the language of the country concerned. In several cases, therefore, it is necessary to resort to more general reviews of European planning or public administration, as is explained in the introduction to the Bibliography.

NOTES: CHAPTER 1

1 See Davies, 1980.

Chapter 2

The Federal Republic of Germany

K. R. Kunzmann

This account of urban and regional planning in Germany represents only one of several possible responses to the question of how much information on the social, economic and cultural structures of the country, or of the historical evolution of planning, is necessary to an understanding of the subject. Unfortunately, literature in English on planning in Germany is limited,[1] possibly simply as a result of the language barrier. On the other hand, it may be the absence of international reputation, of self-consciousness on the part of German planners when disseminating information, or just a certain reluctance to elaborate the basic principles of urban and regional planning in Germany, when day-to-day planning problems have to be tackled first.

Three further points must be made by way of introduction, concerning terminology, territorial definition and the unavoidable antagonism of physical vs economic planning. Terminology is always a problem in cross-cultural studies. The German language, for example, offers a variety of terms describing regional planning on different levels, indicating the subregional, regional, or supraregional. Thus, in order to avoid misleading translations the German terminology is used with a translation which is as near as possible to the usage in British planning practice. Even within the country, terms such as *Raumplanung* are not always used consistently.[2]

A second difficulty arises from the fact that Germany is a federal country. The eleven *Länder*, or states, have slightly different planning systems, and there are for example approaches of different *Landesplanungsgesetze* (state planning laws) besides federal laws, regulations and instruments which are valid for all eleven *Länder*. A compilation of the eleven *Länder* planning systems and the analysis of their differences is far beyond the scope of this chapter. Different political attitudes and day-to-day procedures lead to regional characteristics which do not necessarily reflect the planning systems and approaches of other *Länder*. In general, the system prevailing in the state of Northrhine-

Westphalia forms the basis of this account, but the principles underlying the practice of urban and regional planning and its problems do not vary too widely throughout the country.

Another vital issue is the dichotomy between physical and economic planning. As the goals, tasks and instruments of urban and regional planning should comprise both elements of planning, no distinct differentiation of physical as against economic planning is made. Physical planning procedures and instruments, however, are emphasised when describing the legal and administrative framework of urban and regional planning in Germany. Nevertheless, it is argued that socioeconomic and political constraints and incentives are the decisive factors in the achievement of urban and regional planning goals, which relate primarily to the living and working conditions of the population in a certain area.

EVOLUTION OF PLANNING AND DEVELOPMENT PROBLEMS

The process of building a modern nation state from a mediaeval patchwork of 360 states or free cities was long and difficult, and the functioning of the present federal system represents positive means of achieving this which respect local differences. Local differences are reflected by variations in planning and administrative practice between the *Länder*, and another consequence of late unification is a balanced settlement structure based on formerly independent seats of government or commercial centres.

Three major phases of urban and regional development and policies can be identified. The first is the reconstruction period, from 1946 to around 1959. Problems of recovery from the disastrous impact of the war on German cities and the economy, the partition of the former German Reich and its social and economic consequences, and the inflow of refugees from occupied areas in the East had to be tackled. Democratisation took place under American, British and French control, and a Western capitalist economy developed with the aid of the Marshall Fund.[3]

This was followed by a period of stabilisation during 1960–73 dominated by policies of economic growth and the development of a social system that satisfied aspirations towards an affluent society. It was the period of the *Wirtschaftswunder*, or economic miracle, and issues concerning the quality of life were not allowed to interfere with economic growth. The first indications of discontent came with the students' revolutionary movement of the late 1960s, but it was not until the oil crisis of 1973 that the nation as a whole reacted and began to acknowledge the end of continuous economic growth and consumerism.

By 1974, therefore, the period of stagnation commenced, following the 1973 oil crisis, and it continues still. A new generation of reformist students left the universities advocating qualitative rather than quantitative development. Ecology became an issue, and ecological objections were raised against further exploitation of natural resources. Crisis management, policies of muddling through, dropping out and terrorism came to dominate public discussion. The social and political background of these three periods cannot be analysed in detail here, but they have to be considered in order to understand urban and regional planning policies during the postwar period.

RECONSTRUCTION PERIOD

The first priority in the reconstruction period was the rebuilding of the many devastated cities. In some of them more than 75 per cent of the built-up area had been destroyed during Allied air-raids. Necessarily, the housing sector had first priority. *Sozialer Wohnungsbau*, or the government-subsidised housing programmes for low-income families, and encouragement of housing co-operatives, were the policy answer to the problem of housing more than 8 million refugees from Eastern homelands in addition to the badly affected local inhabitants.

Physical reconstruction of housing was accompanied by the rehabilitation of the infrastructure of transport and water supply, schools and hospitals. Slow economic reconstruction, still related to reparation payments, took place at first. Later those political forces which sought a pro-American capitalist economy rather than a role as a neutral buffer-state gained influence, and economic growth accelerated, utilising Marshall Aid.

The partition of Germany was not envisaged in the immediate postwar years, but became a reality in 1949 when the Federal Republic was created from the territory under Western military control. This led to relocation in the West of a number of industries formerly located in the Eastern zone. Several small country towns, especially in Southern Germany, profited from these firms' need to find suitable locations and qualified manpower to resume production.

Reconstruction of towns and cities generally re-created the prewar land-use pattern. Despite strong arguments for revision of the land-use pattern, and consolidation of plots and ownership, the pattern dating back to mediaeval times was re-established, with only a few new road lines being defined even through those urban areas that had been totally destroyed. Ideas of public ownership of land, to overcome problems arising from private ownership, were rejected, with no more than a reference to the social responsibility that is attached to ownership of property finding its way into the *Grundgesetz*, or Basic Law of the Federal Republic.

An American-inspired anti-planning attitude prevailed, as the democrative alternative to a centrally planned economy. Planning was perceived as interfering with personal choice, and it took some time before any form of spatial planning was again accepted as a necessary means of ensuring satisfactory use of land and resources, and organising social and economic development. However, the influence of the British new towns programme, Abercrombie's plan for Greater London and Swedish experience led to the reintroduction of urban planning. Regional planning had been highly developed under the Nazi regime, and therefore took longer to find a new identity, new goals and respectability. Continuity of administrative officials led to its re-appearance with new political goals and economic instruments. Regional planning was pure physical planning, based on Christaller's central-place theory, and its main objective was the creation of a hierarchy of urban places to obtain a balanced settlement system in the newly created *Länder* throughout Germany. Areas which lagged behind in economic development were designated *Notstandsgebiete* (depressed areas), eligible for federal government aid. Later on, in 1955, most of these areas were renamed *Fördergebiete* (promotion areas), which received considerable government subsidies for economic and social development.

Economic policy and planning at that time emphasised the creation of special incentives for investment in those regions close to the so-called iron-curtain border wth Eastern Europe. Here military and ideological objectives were mixed with social and economic goals.

The provisional selection of Bonn as the new capital of the Federal Republic of Germany and locational competition among the Allies, as well as actions such as the creation of three independent chemical industries out of IG Farben and the brilliant idea of a British army officer to use the remnants of Hitler's and Porsche's Volkswagen works for improved transport for his batallion, which led to the birth of the Volkswagen empire in Germany, are all events which greatly influenced the subsequent distribution of economic activities and social development.

STABILISATION

The reconstruction period was extremely successful, reflecting the feverish reconstruction mentality, the work ethic, the steadfast will to create a better Germany, and the financial and ideological support of the USA. This enthusiastic advance continued unbroken in the second postwar period of stabilisation, when the economic miracle allowed growth to take place in all sectors.

Comprehensive reconstruction was considered to be no longer necessary, and more ambitious goals for urban and regional development were formulated. Planning as such was fostered in order to prepare the ground for new development. Town expansion schemes and the creation of satellite towns at the fringe of all major German conurbations were planned. The concept of the regional city was developed and applied to Hanover, Hamburg and Frankfurt. At first, transport provision dominated planning. Huge road-construction programmes to cater for the rapid increase in car ownership were promoted, with uncritical support from politicians, industry and the public, and pressure from a powerful roads lobby. Advanced planning techniques were introduced from America, which seemed to offer convincing justification for road construction. During this period the legal basis in federal legislation for planning was created.

A new urbanity, to replace the garden-city ideas of the early postwar period, was the goal in urban design. Influenced by Jane Jacobs[4] high density was accepted as the means of regaining the traditional urban atmosphere. Pedestrian streets created in this period gained an international reputation. Another aspect of growth policies at this time was the building or extension of underground and rapid-transit rail systems. These were often based on unrealistic demographic and economic projections, and reflected euphoria on the part of politicians and planners. The system in Munich, built for the Olympic Games of 1972, is a pertinent example. Later reductions in economic growth curbed some of the more ambitious programmes. More buildings of historic importance have been destroyed during this period than during the air-raids in the war. The consequences of this period of euphoria still need to be tackled today.

Intercommunal planning authorities, such as the new Verband Grossraum, Hanover, or the Siedlungsverband Ruhrkohlenbezirk, gained importance. They had to deal with the problems of extensive urban expansion and sprawl at the fringe of the agglomerations and of development pressures upon independent small communities in commuting distance of major urban centres where the majority of jobs and shops were to be found. These self-governing institutions operated successfully outside the existing administrative hierarchy as long as economic and demographic growth seemed to be lasting indefinitely.

A major task in this period was to manage the first symptoms of structural change which affected areas based on a single sector of industry, such as the Ruhr. The development programme of 1966 for the Ruhr – which according to the official view successfully overcame the first economic crisis of the Ruhr – was the political and technocratic answer to the problem. Unfortunately, it was only a single effort, rather than a continuous monitoring and adjustment of policy actions and instruments.

In the latter part of this period shortage of land impeded development, and urban renewal became a priority, with clearance and redevelopment being the prime objectives. Comprehensive urban renewal schemes were prepared from small and medium-sized towns such as Ulm, Tübingen and Ettlingen, seeking to redevelop considerable parts of the town for office and commercial development.

Some very bureaucratic legal procedures for public participation already existed, but these were offputting to the public, and the introduction of improved opportunities was strongly supported by planners and politicians. Planning issues became increasingly prominent in local policy and the media, and urban planning became no longer a technocratic activity.

In a majority of *Länder* urban centres were assigned to become growth poles for regional development based on central place and growth-pole theories. These in turn were connected through a system of development axes, a theoretical concept aiming at concentrating linear infrastructure (roads, rail, and so on) in development corridors.

Planning at the federal level experienced a promising take-off during this period. In accordance with the provisions of the *Bundesraumordnungsgesetz*, the first planning report of the federal government was submitted to Parliament in 1966. This, and subsequent bi-annual reports, contained the results of thorough monitoring, discussed urban and regional development problems, and assessed the outcome of plans and programmes. These reports even included reference to international bilateral co-operation in regional policy. In addition, a federal regional development programme was initiated in 1969. Six years later, after protracted negotiations with the *Länder*, this was submitted to the government. It aimed to set out a long-term spatial strategy for regional development of the whole federal territory.

STAGNATION

The period of stabilisation was succeeded by the stagnation period, which runs from 1973 to the present, when the oil crisis shocked people in Germany and throughout the Western countries. Earlier warnings of recession and unemployment became a reality. The reform programmes of the Brandt era in the 1960s and the related planning euphoria ran into financial difficulties. Many projects and programmes experienced serious reductions and cutbacks. Comprehensive planning, as such, became suspect again. The uncritical belief in planning policies turned to mistrust when planning failures became obvious. Deficiencies were sought in planning, rather than elsewhere, and conservative attitudes were again adopted as a response to the crisis. Planning as a tool for long-range ordering of social and economic activities found decreasing political and financial support. The federal regional development programme, developed with courageous verve in the late 1960s and early 1970s, was when finally accepted by the *Länder* and the Bundestag in

1975, no more than a pale imitation of the original conception. It remains the most recent attempt at integrated federal planning. Today it is not abandoned, but it seems to be forgotten, and day-to-day problems dominate urban and regional planning. Energy, based on coal or atomic power, has become a major issue, overshadowing other goals.

Earlier achievements in environmental protection, integrated urban development planning and public participation are under review, and certain trends may be observed. Awareness of environmental issues has grown considerably, especially among the young. Destruction of environmental quality through industrial development, road construction, coal and atomic-power plants, and so on, is heavily opposed by a large proportion of the population and has led to the formation of citizens' groups protesting against unpopular measures. Economic growth is no longer accepted uncritically and is held to be responsible for social and ecological problems. The citizen of today is more and more conscious of ways of influencing decisions. Many critical young members of society are politically trained and are well aware of the need to campaign actively for the improvement of living and working conditions.

A new regionalism is the consequence of the internationalism of the 1950s and 1960s. Local or regional characteristics have gained a new dimension and a strong movement for the conservation and revitalisation of local or regional heritage can be observed. The strange entente of conservatives and young radical regionalists in this movement is astonishing.

The changing demographic trends, and projections of a declining population, have startled planners and politicians. The consequences are considerable. For example, all regional and local planning programmes in the Ruhrgebiet have to be revised and the investment programmes altered. These programmes were generally based on the assumption of continuously increasing population and economic development. The education sector is an obvious example demonstrating the implications of demographic decline. In the 1960s qualitative and economic arguments led to the construction of large comprehensive schools and large school campuses with up to 2,000 children in high-density, inner-urban residential areas. The migration trend to low-density, low-rise residential areas, the gentrification process in old residential areas, the decreasing number of schoolchildren and some recent aversion to big schools will render some of these obsolete. On the other hand, new residential development areas on the outer fringe of urban agglomerations still suffer from a considerable shortage of schools. The contradiction is obvious: empty schools in the inner core of the agglomeration, but continuing demand for new schools in the suburbs.

International economic competition, rationalisation and the introduction of new technologies have led to a considerable reduction of working places in the traditional industrial sector. Traditional locational advantages are no longer so beneficial, and traditional industrial areas such as the Ruhr decline in importance as industrial closures take effect. Huge areas of industrial land within the urban areas are vacant or underutilised, while new industrial land is being developed along the regional motorways, eating up agricultural land and spoiling the natural environment. New conflicts can be anticipated at the outer fringes of the urban areas. The means of controlling this locally are not always available or not used for political reasons. The bitter competitition among the cities for additional trade-tax income makes any concerted approach or action impossible. Thus, the structural change in the economic sector, favoured by general industrial concentration processes, will continue to change urban land-use patterns in and around agglomerations in Germany.

On the other hand, there are hopeful signs that the general economic change will have some positive results. The concentration of shopping activities in city centres, where the attractive pedestrian precincts have had a negative impact on the urban land market elsewhere, seems to have come to a halt, and the unexpected penetration of huge hypermarkets into industrial areas seems finally to be under control, due to strong regional restrictive regulations. These hypermarkets have had disastrous effects on urban shopping centres and corner shops. The concentration of new office blocks along the major highways, adding new congestion to the urban motorway system, also seems to have reached saturation point, and some decentralisation of offices can be observed.

Economic recession leads to further intercommunal competition for new jobs. As new jobs are created this process leads to further concentration in the attractive Rhine corridor and the modern service cities in the south. Rehabilitation and modernisation of the housing stock, combined with measures for traffic abatement in the cities, have become the prevailing goal and task of urban planning policies. Attention has been drawn to the disastrous housing situation only by the squatting movement in the 1980s. Shortages had been forecast long before, but politicians did not react in time. The traditional mixture of residential and industrial land uses seems to be valued again, after years devoted to following the principles of the Charter of Athens by pursuing the objective of rigid separation of land uses.

GOVERNMENTAL CONTEXT

Three major factors dominate the policy and implementation processes in Germany. These are the federal system and its underlying philosophy;

the principle of civic liberty and local self-government; and the presumption in favour of private enterprise rather than state intervention. The basic feature of the federal system in Germany is the assignment of legislative functions predominantly to the federal government itself and the executive functions to the *Länder*. This power of the federal government is confined to legislation having effect throughout Germany.

The Basic Law of 1949, which acts as the constitution of the federal republic, gives only limited powers to the federal government, but subsequent economic and social development has had the effect of giving it greater powers. The states, or *Länder*, have nevertheless retained important legislative fields. These include local government law, aspects of environmental protection (a competency currently very much in dispute) and education, including university education. Education is also subject to federal law. The states have responsibility for implementing and administering legislation, largely free from detailed direction from the federal government.

Responsibility for urban and regional planning issues in Germany on the federal level lies with the Bundesministerium für Raumordnung, Bauwesen und Städtebau (Federal Ministry for Regional Planning, Building and Urban Development). This responsibility includes preparation and continuous monitoring of urban and regional and federal development goals and policies, in order to achieve a balanced spatial development of the federal territory as a whole.

On the *Länder* level there is a co-ordinated and comprehensive system of planning for the development of the *Land* territory. The responsibility of the ministry concerned, for example, the Ministerium für Landes- und Stadtentwicklung in Northrine-Westphalia, is the preparation and continuous monitoring of *Länder* planning laws and development, and supervision of the local planning system.

Local government has a long and strong tradition, going back to the privileges of the free towns in the Middle Ages. Modern local government owes much to the Municipal Order of 1808 and other reforms of the Prussian minister Freiherr von Stein. The Basic Law stipulates that the states must guarantee to local government the right to regulate its own affairs, and that local authorities must be democratically elected.

Local authorities or *Gemeinden* are responsible for local public transport, local road construction, electricity, water and gas supplies, housing construction, building and maintenance of elementary and secondary schools, theatres and museums, hospitals, sports facilities and public baths, adult education and youth welfare, subject only to legal control by the *Land*. Many of these tasks are beyond the resources of smaller authorities and are taken over by the next higher level of local government, the *Kreis*. The *Kreis* is also responsible for a landscape plan. Thus, there is a four-tiered hierarchy of planning to correspond with the four levels of government: federal (*Bund*), *Land, Kreis* and municipal (*Stadt* or *Gemeinde*).

The decentralised system of government is reflected by a decentralised tax system. All three levels, the federal government, the *Länder* and the communities have their own sources of revenue. The federal government controls about 50 per cent of the entire tax revenue. The other 50 per cent is distributed between the *Länder* and local communities, which receive the full yield of the land tax, 60 per cent of manufacturing tax and 14 per cent of income tax. This revenue is nevertheless normally insufficient, and local authorities frequently need to seek grants from the *Land*. The system of redistributing revenue to compensate for differences between prosperous and poor areas is often admired, but it is a regular source of political dispute. This is especially true for the local trade tax which is a major source of intercommunal conflicts and competition to attract industry to communities through zoning, the development of industrial land and industrial promotion divisions in the local authorities. Financial assistance from the *Länder* is often tied to specific purposes such as school-building, social facilities, or urban renewal programmes, and can include assistance with the cost of hiring planning consultants to prepare local plans. Subsidies are given according to criteria which tend to favour regional rather than local interests. Extreme disparities between districts is avoided, but local initiatives may be stifled.

The general principle of private initiative rather than state intervention determines the way in which instruments for economic development and social policies are developed, as is obvious in the housing sector. In the absence of public housing schemes financed and owned by local authorities, private housing for low-income groups, including housing owned by co-operatives, is provided by private initiative and subsidised heavily. This principle is also adopted for industrial promotion, where tax incentives, subsidies for land development, training, research and a variety of other forms of financial aid are given to private entrepreneurs, upon application. The support of private initiative is an essential element of social and economic development policies.

PLANNING LAWS AND REGULATIONS

German planning legislation has a remarkably long tradition. Apart from building regulations which were applied to the development of towns since the Middle Ages, the first planning law in the proper sense was the Building Lines Act 1868 in Baden. This Act was followed by similar laws and regulations in other regions. In the following decades more and more need was felt to regulate urban and regional development. Zoning ordinances were formulated and enforced. In the years 1933–45 a so-called *Reichsstädtebaugesetz* (Imperial Town Planning Law) was prepared, but never completed and enacted.

The modern era of planning legislation was established when, in 1960, the *Bundesbaugesetz* (Federal Planning Act) was passed after ten years of discussions between the federal and *Land* parliaments. This Act, with its subsequent amendments of 1969 and 1976, regulates urban planning and assigns it to the *Gemeinden*. It provides the legal framework for the *Flächennutzungsplan* (land-use plan) and *Bebauungsplan* (building and zoning plan), and regulates the implementation of these plans with regard to building control, land acquisition and land markets, consolidation of property and compulsory purchase. This Act is complemented by *Land* legislation. Responsibilities and principles of regional planning are defined in a federal skeleton law, the *Bundesraumordnungsgesetz* of 1965, amended in 1976.

Housing and urban renewal was promoted by the *Städtebauförderungsgesetz* (Urban Development Act) of 1971, which sought to facilitate urban renewal and expansion. Through the requirement for thorough social surveys and extended public participation in the development process this law protected urban citizens against pure capital interests in redevelopment or town expansion. A related measure at federal level is the *Wohnmodernisierungsgesetz* (Housing Improvement Act) of 1977, concerning the various loans and subsidies available to propertyowners for modernisation and renovation of old private housing stock.

During the reconstruction period there was a lack of legal planning regulations. As a consequence, most of the *Länder* adopted reconstruction legislation very quickly around 1949, most of which was based on prewar ideas adapted to postwar priorities. Northrhine-Westphalia, for instance, passed a Planning Act in 1950. With amendments in 1962 and 1975, this law defines the organisation and goals of the *Länder* and regional planning, and the way in which the objectives mentioned are to be implemented, and includes provision for a *Land* development programme.

Additional planning-related legal instruments in Northrhine-Westphalia include regulations concerning noise control in urban areas; the design, protection and conservation of townscapes and landscapes; regulations on the consideration of environmental issues in planning; the design of playgrounds; location and control of camping-sites; protection of woodland; residential densities and settlement foci; the location and control of hypermarkets; and land consolidation. One controversial measure was to combat urban pollution by controlling the minimum distances between specific land uses.

Elsewhere in Germany an interesting procedure operates in all *Länder* except Hessen and Northrhine-Westphalia. This is for a kind of public inquiry (*Raumordnungsverfahren*) in which public authorities, but not private citizens, participate in a decision as a major project such as a new airport or motorway, seeking to reconcile their interests. At the local level, under Northrhine-Westphalia law, every *Gemeinde* has the right to pass an *Ortssatzung*, or local ordinance, to protect historic buildings and the local scene from visual intrusion.

PLANNING AS A PROFESSION

The planning profession in Germany has evolved along two main routes. At a local level architects responsible for housing matters gradually qualified on the job as planners. They still hold a majority of positions at the local level, working in local authority planning offices. On the other hand, planners with qualifications as lawyers, geographers and economists hold a majority of posts on regional, *Länder* and federal level. Since 1975 planners trained as undergraduates in urban and regional planning have entered professional practice and compete with graduates from traditional disciplines.

In most cases planners in Germany hold permanent positions as civil servants for which they qualify only after a two and a half year training programme with various planning departments on different levels. Chief planning officers of a town are elected for a fixed period of between six and twelve years. Generally they have to be a member of a political party. Thus, they hold a political position, with all the positive and negative implications that this has. Towns and cities normally employ their own qualified planning staff, but small rural communities frequently have to employ consultants instead.

Local planning practice used to be concerned primarily with plan-making and urban design, but present-day work consists much more of preparing short-term political decisions, of reviewing existing plans, monitoring development changes, and preparing and performing information activities in the context of public participation procedures. Regional planners at the different levels perform, in contrast, a more administrative role. Nevertheless, their sociopolitical creativity is no less important in their controversies wth politicians and planners from the communal level.

The Federal Republic of Germany does not possess a strong professional planning institute. The Bund Deutscher Architekten (BDA) (Association of German Architects) is still the only powerful institution for architect-planners (non-architect planners are not eligible for membership). This is also true for the chambers of architects. Their influence on the adherence of the physical planning approach to planning and planning education is considerable. In the early 1970s a specific professional organisation for urban and regional planners evolved, the Vereinigung der Stadt-, Regional- und Landesplaner (SRL) (Association of Urban and Regional Planners), which is open to planners from all planning-related disciplines. Two institutes whose membership is restricted have to be mentioned in this context: the Deutsche Akademie für Städtebau and Landesplanung (German Academy of Urban and Regional Planning) and the Deutsche Akademie für Raumforschung und Raumordnung (German Academy of Regional Spatial Research and Spatial Policy). Their membership includes senior planners from academic life and planning practice.

All these institutions have only limited influence on official urban planning programmes at universities. Their advice, however, is continuously sought by those ministerial commissions which are responsible for the approval of curricula and examinations. The official approval of the civil service examination board, on the other hand, is a vital form of recognition of university degrees in town planning.

During the stabilisation period the influence of British and American experience stimulated the emergence of planning education in its own right, quite separate from architectural education. As a consequence, a variety of routes to qualification in planning now exists. Comprehensive undergraduate programmes are offered at the University of Dortmund and the Technical University of Berlin. Both favour a generalist approach, combining social, political, economic and technical aspects of the subject in a four and a half to five-year course. A similar approach is being adopted at the University of Oldenburg and, with more physical emphasis, the University of Kaiserslautern.

A planning specialisation within an architecture curriculum is the traditional and most frequent form of planning education, offered at every department of architecture in the technical universities. It is most highly developed in Aachen, Munich and Stuttgart. Postgraduate programmes exist in Munich, for architects and engineers, and in Karlsruhe, where the regional science course is available to a wide range of graduates interested in theoretical and methodological knowledge of planning issues.

Planning is taught within the curriculum for several other degrees, but these are not recognised for entry to the civil service training programme for planners. Consequently, such graduates are restricted, in the urban planning sector, to looking for jobs in major authorities which have established urban research departments. They can, however, take regional planning posts.

The crucial question regarding the need for planners is difficult to answer. Due to government education policy, which has been aimed at the quantitative expansion of university education, there is no clear interdependency between the demand for urban planners and the university places provided for training. This has resulted in recent years in a slight overprovision of planners. Whereas in the 1960s the enormous demand for urban planners led to a vast intake of planning students which could be financed due to unbroken economic wealth, the young graduate today is faced by the fact that many posts are occupied by relatively young professionals and that new positions can hardly be financed even if there is a clear social demand.

PLANS AND PROGRAMMES

The hierarchy of planning and programming institutions as ideally conceptualised and legally based has already been described, and now the scope and contents of individual plans and programmes can be briefly reviewed. In contrast to the sequence used above, the following section starts with the lowest level in the hierarchy, the local level of planning, and proceeds up to the federal level. This is deliberate, to demonstrate that local plans are not derived from federal plans, but from local conditions, resources and demands of the population concerned. It may, however, equally demonstrate the feedback processes which characterise the German planning system. This sequence from the local level, where impressive and increasing citizen involvement takes place, to the federal level, where only a few 'representative' politicians decide upon policies, laws and programmes, illustrates the range of opportunities to influence decision-making processes.

On the local level there are two types of legally based plans: the *Flächennutzungsplan* (land-use plan) and the *Bebauungsplan* (zoning plan). The process of setting up these plans is called *Bauleitungsplanung*. The land-use plan is a preparatory plan, setting the context of planning policy for a district, within which the legally binding zoning plan may be prepared for a specific precinct.

The *Bebauungsplan* (zoning plan) shows in detail the delimitation of public and private land, the kind and density of land use, built-up land and land which has to be kept open. Depending on local conditions and requirements more detailed regulations may be indicated, for example, the kind of roofs to be used, the planting of trees, or the arrangement of parking-lots. As a rule, the *Bebauungsplan* has to be developed out of the *Flächennutzungsplan* – but often partial plans passed long before a *Flächennutzungsplan* covering the whole community has been approved. The *Bebauungsplan*, being a local ordinance, is legally binding for every citizen and public institution without exception. It is the legal precondition for the implementation of urban planning and all construction takes place on the basis of its provisions. Normally it is on a scale of 1:1,000, but it may need to be at 1:500, and is accompanied by a written explanation of the policy.

The local authority must arrange public hearings or opportunities for public participation in association with the plans. Whenever it might have a serious impact on the living and working conditions of the local population, a *Sozialplan* (social plan or programme) has to be prepared in co-operation with the citizens in the area affected in order to minimise the adverse effects.

The *Flächennutzungsplan* (land-use plan) is a land-use plan for the whole territory of a community in the scale of 1:10,000 or of 1:20,000 depending on the overall area of the town. This plan is not legally binding, but does bind all public institutions as far as their local land-use allocation is concerned, and gives general guidance for overall urban development. In principle this plan is revised every seven to ten years, but in practice small revisions are made more frequently in order to give permission to specific projects. In association with the land-use plan is a complex system of participation with organisations which have certain public responsibilities such as public utilities, churches and chambers of commerce in order to achieve a degree of sectoral planning. Once approved, the plan becomes binding on these institutions.

Urban renewal and new housing projects are carried out under quite separate legislation, the *Städtebauförderungsgesetz* of 1971. Essential elements of the process include extensive surveys to determine the need for rehabilitation or new building, and identify attitudes and response of the residents, social consequences of the policy and appropriate remedial action. A social plan or programme is prepared to indicate all of the necessary social measures, and the physical development is defined in a *Bebauungsplan*. Requirements for public participation and involvement are more strict than in the usual planning procedures elsewhere. Financial aid may be sought from the *Land* government, and outside organisations may be appointed to manage the whole process within a designated renewal area.

Other planning instruments are sometimes used, particularly the *Stadtentwicklungsplan* or urban-development plan. These have no statutory basis, but are concerned with co-ordination of sectoral plans for education, transport, and so on, and programming financial and legal provision for development, as well as land-use allocation. In several cities policy departments for corporate planning and urban research have been introduced for this purpose.

A comprehensive approach to urban planning, by this means, was pioneered in the late 1960s with great enthusiasm. However, the expectations held for integrated urban planning proved overambitious, and more pragmatic and *ad hoc* approaches are now to be found. Many worthwhile elements remain, but the failure of comprehensive planning has caused much frustration and is to be regretted.

The eleven *Länder* of the federal republic regulate their regional planning independently, within the framework of federal legislation, the

Bundesraumordnungsgesetz. As an example, the plans and programmes in Northrhine-Westphalia are described here. Spatial development is guided, at the *Land* level, by the land-development programme (*Landesentwicklungsprogramm*) and land- and area-development plans. Sectoral landscape plans (*Landschaftspläne*) are increasingly important in the undeveloped areas. In addition, special development programmes have been formulated, and a bi-annual report on regional development monitoring and assessing the effectiveness of plans and policies is published by the office of the minister-president of the *Land*. These reports provide a valuable overall review of regional development.

The first land-development programme was approved in 1964. This sought to concentrate development in certain growth poles and axes of development, in order to create a more efficient structure and protect the rural areas from urban sprawl. In 1974 an updated version was approved, which extended the scope and technical sophistication of the earlier version, but retained its general principles. The first programme assumed a growing population, and the second a static population until 1985, but in fact the population is now decreasing. These programmes represented the government's goals for physical development and are not detailed budgetary programmes, although expenditure decisions are closely related to them.

A total of six land-development plans (*Landesentwicklungspläne*) are proposed. They have been prepared with wide consultation with all institutions and interest groups affected and are concerned with specific sectors of policy. Plans I and II, approved in 1979 after considerable political lobbying, concern spatial and settlement structure. Revisions are now being considered in order to adopt more satisfactory functional areas as spatial units for policy-making. Plan III, approved in 1976, relates to open space, water management and recreation.

Plan IV is concerned with noise emission from airports and noise-protection areas. This is still a draft plan, possibly due to the sensitive nature of airport development issues. Plan V, relating to mining and mineral-resource development, is also politically sensitive, and is being prepared in some secrecy. Plan VI, approved in 1978, concerns the allocation of sites for major land-consuming projects including refineries, coal extraction, and coal and atomic power plants. The effectiveness of this plan is affected by changing political and technical attitudes.

These six plans represent a major achievement in creating physical development policies for such a large and densely populated agglomeration. They are interpreted in more detailed physical planning terms by a series of subregional structure plans (*Gebietsentwicklungspläne*), which in turn co-ordinate local development plans. These are land-use plans at a scale of 1:50,000. They are criticised as being purely physical planning, and for limiting the autonomy of municipal authorities in determining such issues as residential density and land use.

An important new means of protecting the natural environment and ecologically sensitive areas is the *Landschaftsplan*, or landscape plan. This is based on a law passed by the *Land* Parliament in 1975. Economic development programmes have been prepared from time to time, often as a political response to economic problems. Programmes were prepared for the Ruhr in 1967–8 and for the whole *Land* in 1975, and currently there is the *Aktionsprogramm Ruhr*, 1980–4. This programme includes measures for job creation, research and development, environment protection and improvement of residential environment, energy resources, investment and cultural development.

It is hoped that great improvements will be achieved by these measures, but all the various planning measures are a source of political dispute concerning state intervention into economic processes. The need for co-ordination of policies, guidance of development and allocation of land uses is, however, generally accepted.

At the federal level it was some time before any national planning became acceptable. In 1969 a federal regional programme was proposed, and eventually in 1974 a programme acceptable to Parliament was passed. The overall goal of the programme is to achieve balanced regional development and equal living and working conditions.

There is little prospect of any form of national development plan, but there are at the national level sectoral plans for transport and energy, and regular co-ordination conferences between the appropriate *Länder* ministers. The federal government has published since 1966 a bi-annual report on planning and development, and at the national level an advisory council exists to give expert technical advice to the government on regional planning.

In the period since the first (and possibly only) federal regional programme was approved achievements attributable to it have been limited, and expectations disappointed. Regional planning now has less political support, and economic difficulties coupled with limits to the extent to which co-ordination of policies can be achieved, contributes to this situation.

CONCLUSIONS

In addition to describing the German planning system, several criticisms have been offered, but the system certainly has produced some successes. The building-control system; the provisions for citizen participation; and the complex hierarchy of plans, programmes and administrative links are all worthwile features. Also the creation of nearly 100 pedestrianised city centres is internationally recognised to have contributed to the restoration of historic city centres and the reduction of some inner city problems.

Internationally the most interesting feature is probably the system of plans and programmes at *Land* level, with its procedures and powers, and public and political control. The relationship between physical and economic planning is still not satisfactory, but the balance of public and private interests is as a rule quite successful.

Public participation is now generally accepted, in spite of lengthening decision-making procedures. Younger people are increasingly aware of planning issues, which are often the subject of school lessons especially in the bigger conurbations.

The relative lack of imbalance in the settlement structure or of disparities of wealth may be cited as planning achievements, but these may equally be attributable to federalism and local government autonomy, and their historic roots. However, the even distribution today of urban centres, and the role of smaller towns and cities in the urban system, is remarkable. The final evaluation of the performance of German urban and regional planning must nevertheless be left to planners from elsewhere, to judge from a comparative standpoint.

NOTES: CHAPTER 2

1 There is only one up-to-date bibliography of source material in the English language on urban and regional planning in Germany available, which may help those interested in more information on the issues raised in this chapter. This bibliography (Bach, 1980) is a revised and updated version of a bibliography that the author initiated and published in 1975: W. Coprian and K. R. Kunzmann, *English Language Literature on Urban and Regional Planning in the Federal Republic of Germany 1955-1975*, Dortmund, University of Dortmund, 1975.
2 The term *Raumplanung* (spatial planning) in principle covers space-related planning at all levels from integrated, comprehensive urban planning up to federal development planning, but is also used for regional planning only.
3 See, for example, Crawley, 1973; Childs and Johnson, 1981.
4 See J. Jacobs, *The Death and Life of Great American Cities*, Harmondsworth, Penguin, 1964.

Chapter 3

Italy

S. Bardazzi

Planning at both urban and regional scale is reviewed here, but attention is given primarily to town planning. Regional planning is discussed later, and in less detail. The form of town planning at the municipal level has not been modified greatly since 1942, although aspects of the system are now being critically debated. On the other hand, effective regional planning, and the associated institutional organisation, is still in its infancy.

Although provision for regional government was made in the constitution of the republic promulgated in 1947, regions started to function in their present form only in 1972, and the consequential legislation to transfer power has not been fully implemented. This has important implications for planning. Not only are the institutions of regional planning therefore not fully operational, but aspects of urban planning and decisions on roads, aqueducts and other public infrastructure are held up by the regions.

Town planning, in the strict sense of the term, has been the principal form of planning practised since the war. During the late 1970s the concept of town planning has received a wider intepretation than that embodied in the basic planning law of 1942, although the institutional arrangements have not been substantially modified. In spite of this trend, the concepts behind the 1942 law remain a point of reference for present legislation being proposed by the regional authorities.

The culture of city life has had a great influence on institutional organisation, and ensured a favourable position for urban rather than rural interests. Even the major development schemes, intended in the postwar period to be prepared for all land, have never become particularly significant. The importance of the towns, both as a residential environment and in a political sense, prevails.

The development of regional science, or the analysis of data concerning overall spatial organisation of the land, is fundamentally lacking. Urban planning has concentrated instead on building development, especially in the context of the growth of existing centres. Implementation of public works and transport facilities by sectoral agencies has also been very influential over urban land use. Rural areas, in contrast, have generally not been regarded as a problem for planning.

There have been attempts to reform urban planning, and solve the problems of particular parts of Italy such as the south, or sectoral problems such as motorways, aqueducts and ports, but a general strategic policy for planning has been totally lacking, with obvious consequences.

Nowadays the regions are attempting to organise their territory using more sophisticated techniques of planning, and national legislation is giving particular attention to the existing housing stock. Historic town centres are no longer being treated as monuments and are subject to policies of restoration. Recent regulations in 1978 have led to plans concentrating on individual buildings, although both ecological and historical interests advise the protection of the environment and its cultural values, and economic considerations would also suggest a wider view, combining restoration and renewal.

The following discussion will be devoted chiefly to the planning processes surrounding layout plans. As has been indicated above, the emphasis and order of topics derives from political characteristics of the Italian culture. First, however, brief mention should be made of the constitutional framework. The constitution of the republic was promulgated on 27 December 1947 and came into force on 1 January 1948. It states that the republic is divided into regions, provinces and communes. Five regions, Sicily, Sardinia, Trentino-Alto-Adige, Friuli-Venezia-Guilia and Valle d'Aosta, are assigned a degree of autonomy by special constitutional statutes, and are therefore somewhat distinct from the other fifteen regions. Law no. 281 of 1970 initiated the transfer of power to the regions, and regional statutes were approved during 1970. As far as planning is concerned, the communes or municipal authorities and the regions are the important authorities, and the provinces of much less significance.

PROMOTERS OF PLANNING

These can be categorised as public and private operators. On the public side, the commune or lowest level of local authority is most important. The basic planning act, law no. 1150 of 1942, allocates the commune a major role in planning, and this role was maintained by the postwar constitution. It requires the commune to prepare a communal masterplan known as a *Piano Regolatore Generale Communale* (PRGC). The law also allows national authorities to prepare a development scheme to co-ordinate planning work to be carried out locally, but such schemes do not lead to a hierarchy of plans or involve the regions or provinces. A development scheme is called a *Piano Territoriale di Coordinamente* (PTC).

One can see in the 1942 law an authoritarian model of planning, since the approval of all plans depends on national government. Nevertheless, it allows the communal (PRGC) level of plan to be implemented in the absence of any PTC plan. Every local plan can rely on a certain degree of autonomy, subject to co-ordination only in certain areas of the country, and where a PTC is prepared before a PRGC.

Circumstances of war did not allow experimentation with law 1150, so we lack evidence to support a judgement of how it would have been operated by the government which promoted it. The immediate aftermath of war saw plans for rebuilding war-damaged areas, under laws of 1945 and 1951. These were related to the initiatives designed to restructure the economy with international aid. Reconstruction plans were prepared by the communal authority. They did not cover the whole district, but only those areas having damaged building. These plans had the effect of re-creating the previously existing layout, although this was not the explicit objective as it had been with the prewar alignment plans or building-regulation plans. This mode of planning for reconstruction was an obvious response to the need to repair war damage, but it was also a most restrictive form of planning. The PRGC would have been a valuable planning tool in this situation, but the rebuilding plan was appropriate for the kind of development sought at the time, and also reinforced the connection between property development and profit. During the period 1945–54 the commune acted as a promoter of development, as it did later in different ways, but normally it did not undertake development directly itself.

Postwar conditions were difficult for all local authorities but were particularly severe for the communes. In addition, law 1150 had assigned them tasks which they were not willing to carry out due to the lack of the necessary administrative structure, expertise and resources to operate as a planning authority. The situation encouraged private initiative, since the land was largely privately owned, and developers with the right to build (*jus aedificandi*) could obtain considerable credit facilities. The commune drafted the plan, but implementation was in the hands of private enterprise, which showed ingenuity in avoiding control, or worse, operating illegally. During this period the commune plans directed reconstruction, generally recreating the former pattern of development, but the extent of the built-up area was extended by various more or less legal contrivances so as to make reconstruction as economical as possible. One may conclude, however, that the principal operator in urban planning was the private sector.

In 1954 the period for reconstruction plans formally ended, and a list of communes required to prepare a PRGC was proposed. The requirement to prepare a plan gradually replaced the advice to do so, underlining the legal role of the commune in the planning process, but the communes remained ill-equipped to prepare plans, and prospects for communal planning were if anything worse than before. A deep gap was noticeable (and still is today) between what the commune should do and what actually took place.

When a PRGC was prepared covering the whole area of a commune, it took the form of a programme for development and the use of land, and determined the extent and distribution of building in the different homogeneous areas of the commune. It did not attempt to plan for productive sectors of the economy and their territorial relationships. The same approach characterised the intercommunal plans which do not form a higher level of planning, but exist to co-ordinate development in adjoining areas of two or more neighbouring communes.

In 1967 a so-called Bridge Law gave a new impetus to planning and assigned a primary role to the commune to regulate private building activity. On the other hand, the semi-legal practice of private-plot subdivision plans became legalised as communal implementation programmes. Urban planning standards were adopted for control of buildings as a compulsory and standardised part of communal planning control under this law. Previously standards varied widely due to different planners' views and the discretion of the administrators. An earlier law, in 1962, which gave the commune powers to purchase land for development of low-cost housing, as well as the 1967 law, tended to promote the role of the commune as a planning authority. However, lack of financial resources have limited the extent to which positive results have been achieved. Under the law of 1967, the control of private initiative through the expensive negotiation of subdivision plans makes the commune a public operator with greater power of decision than the private developer, who sees the *jus aedificandi*, which was once decisive and unquestionable, lose meaning, and his speculative opportunities diminish.

THE COMMUNE

Thus, it is argued that communes have acquired responsibility for planning, and a prevailing position in the planning process, so they should be discussed further. There are 8,096 communes in Italy, varying in population from a few dozen to several million. Individual communes often include wide variations of socioeconomic status within their area. Communes have more formal than substantial powers, and are governed by an elected council. Voting is often influenced by attitudes to national political issues, and a local council is often controlled by parties in opposition at a national level, or partially so in a more complex coalition. Planning powers are assigned to all communes on an equal basis, but clearly there is a wide variety of planning issues faced by the communes, and of ability to tackle them. Although planning powers of the communes have been increased, their structure and resources have not been strengthened in parallel. Transfer of power from regions to communes only started in 1977, and in spite of many years' debate on local authority autonomy, they are still incorporated on the basis of old and inadequate laws.

The different forms of urban planning that evolved in the postwar period up to 1967 were characterised by a design or blueprint ideology, limiting their practical applicability. At the same time, debates on urban planning centred on such issues as the nature of plans, planning competence, reform of legislation, and links between urban planning and environmental conservation or economic planning. It is paradoxical, in view of these considerations and of the nature of the communes, that reforms of the planning system should take the form of strengthening the role of the communes, as happened with the reforms of 1967 and new laws in 1971.

In 1971 communes were given greater powers of expropriation of land in a five-year programme, to implement the PRGC. The latter was also required to be co-ordinated with a new sectoral plan for commerce. This link between a sectoral plan for distribution and trade and the PRGC was confined to land-use allocation. Consequently, the potential benefits, particularly of a thorough study of the commercial sector in order to make decisions about commercial development linked to land-use decisions, were not fully realised.

Another measure passed in 1971 is of particular interest, although it applies only to designated mountain communes. This provides for the preparation of an economic development plan alongside an inter-communal land-use plan similar to a PRGC. Mountain areas are designated by the regional authorities and generally consist of those communes over 600 metres above sea-level. These mountain communes, occupying over half the area of the country, have the benefit of considerable innovation in planning practice, particularly as a result of this link between economic and physical planning. A group of communes may prepare an economic plan or scheme for the development of productive sectors of the economy, in association with the land-use plan. This procedure is of great significance for the development of planning thought, although it is limited by virtue of its application only to the mountain settlements, and because sufficient financial resources for effective development in underdeveloped problem areas are not normally available. Nevertheless, it represents an approach to planning far removed from a simple urban-development scheme, showing that the PRGC and intercommunal plans are capable of greatly varied application.

Since the late 1960s the issue of the timescale for implementation of plans has been gaining importance. In the basic legislation only the overall time for implementation was included, but this represented the formal period of the plan rather than a realistic programme for implementation. There have been many examples of public authorities being unable to fulfil plans within the stated time either because subsidies were not forthcoming or as a result of slow bureaucratic processes. A deep gulf between development plans and economic or budgetary programmes used to be evident, since the PRGC had, in principle, a static character, while the communes also prepared development schemes, which were in effect corporate plans for all public investment to be undertaken within one year. Clearly, for the latter, time is a vital dimension.

The requirement to define the timescale of a PRGC came from a decision of the Constitutional Court in 1968 taken to protect the interests of private owners of land, who could suffer blight and insecurity in the absence of any such regulation. Further emphasis on programming came in the light of experience of operating the 1971 innovation on expropriation procedure, with the introduction of an implementation programme covering three to five years in law 10 of 1977. This implementation programme is prepared by the commune administration and is based on a survey of the existing housing stock, opportunities of reuse of older housing and programming of new settlements. In addition, this law finally broke the link whereby the right to build derived from the rights inherent in ownership of property. The power of communes increased, as the right to build had henceforth to be granted by the mayor.

In considering the question of who plans, the two major actors have been identified as the communes and private developers. The changes discussed above strengthened the powers of the communes very greatly, but although the formal position of private developers has weakened, they continue to play a major promotional role in planning and can be expected to continue to find ways of participating in planning decision-making. The power of the communes is limited, however, by the powers and influence of national and regional authorities above them in the hierarchy. Some individual communes are willing and able to tackle their own urban problems, but often they have to operate within a framework of an intercommunal or regional plan. Devolution of power to regions has made this less rigid than it might otherwise be. Basic data for planning is compiled for regional policies, and one could perhaps say that a new cyclic style of planning is evident, where information and decisions interact with each other. Uncertainties and contradictions remain, however.

Preparation of plans requires a combination of technical and political contributions. The technical work is undertaken by professionals employed by research or project institutes or communal and intercommunal technical departments. Political power is exercised at all levels of government, and the influence of local politicians – especially that of the ruling group or *giunta* of full-time political leaders of an authority – on the planning process can be considerable.

Democratic participation also plays a considerable role. The law of 1942 recognised the right of certain interests to participate in the planning process, but nowadays it is recognised that all citizens in a community have the right to suggest improvements to a PRGC. The present trend is for public participation to become an increasingly important element in the process of preparing and implementing plans.

THE CULTURE OF URBAN PLANNING

In a country such as Italy – with its rich historical heritage evident in every locality – one might expect that urban reconstruction would follow the urban pattern already established but, in fact, the period of postwar reconstruction (1946–54) is characterised by an untidy pattern of building, mostly the result of speculative development. The culture of urban planning at this time could be described as pre-rationalist.

The main objective at this time was to repair war damage, rather than initiate new urban centres, and as most people engaged in this activity were landlords and private developers, proposals were designed to obtain the maximum financial profit. This is a trivial purpose, clearly in conflict with major social interests, but it was widely accepted. The reconstruction plans followed closely the principles of the building plans established in 1865, in the early years of a united Italy, but the hygiene and design criteria of that period were completely ignored. This form of plan, with a minimum of constraint, but which tended to re-create the old street pattern, was a major factor in the style of development of this period.

Uncontrolled urban growth, which became common in the 1950s, and was described as 'oil-slick' development, led to a movement for a more co-ordinated and coherent form of development. Urban sprawl was a clear consequence of the reconstruction process and the economics of development. In response to this a new movement attempted to create organic communities with a co-ordinated provision of residential development, services and facilities geared to people's needs. This movement was related to the modern movement in politics, and of the community movement, whose leader was A. Ollivetti, president in the 1960s of the Institute of Urban Planners. A number of projects were designed adopting these principles of co-ordination, and reflecting foreign influences, but not in sufficient numbers, or in sufficiently prominent locations, to become a real alternative mode of planning.

By 1955 the period of reconstruction ended, at least in a formal sense, as the PRGC became obligatory for all cities listed in a decree from the Ministry of Public Works (Decree 391, 1954). Rational planning methods were adopted, in principle, and advocated by the ministry in 1954. Proposals incorporated in the PRGC were to be based on a scientific analysis of relevant data, and included plans for the road network as well as land-use allocation. However, the process was not in practice always rational. Proposals were comprehensive and formalised, and not confined to planning considerations. The plan more often reflected an image of the city derived from the planner's culture than the product of objective analysis and serious research. Land-use allocation is, thus, reduced to a distribution of land to the building industry. The tendency remained for development to be carried out by private enterprise in ways which attained the highest profit, and which produced the much-criticised 'oil-slick' development. The PRGC was during the 1950s reduced to the status of a formality which did not alter the traditional methods of the building industry.

In 1962 new legislation attempted to give to the communes the means to carry out promotional planning activity, and not merely regulatory plans. Law 167 of 1962 provided new procedures for expropriation of land to be used for public projects. Unfortunately, the necessary financial resources were not available, and in most cases these schemes became regulatory plans for private enterprise and raised the price of the land.

A principal objective of urban planning is to achieve for the commune effective control of private initiative and ensure that private developers, who are the effective implementers of plans, give due regard to social values. The law of 1962 did not fully achieve this, and efforts continued to be made to reduce the power and the arrogance of private enterprise by stricter methods of land expropriation and refined methods of rational planning.

Law 865 of 1971 sought the first of these by establishing agricultural value of land as the basis for expropriation, with precise reference to the agricultural productivity of the land. The second was sought by adopting a procedure of land-use zoning, and making controls more rigid.

By the early 1970s planning had become an institutionalised feature of city culture, with a basically rational methodology, but also adopting organic concepts of urban growth taking place in complete forms, with residences, services and communications developed in a balanced and integrated manner. Unfortunately, results are still disappointing, because the private sector still dominates the activities of the communes and, therefore, the management of plans.

During the 1960s ideas of regional and subregional planning were debated widely, having already been raised by the Institute of Urban

Planners. Ideas of regional cities and local centres, and of the relation between urban and land-use planning and economic development programmes, were developed, but remained theoretical expressions not put to practical effect. Attempts were made to introduce concepts of national urban structure and organisation, and a co-ordinated solution to problems of regional imbalance. The centre-left government was receptive to legislative initiatives for reform of urban and regional planning, but these never achieved the objectives intended. Institutes of planning studies and regional research have grown and ideas developed, but it is necessary to conclude that the gulf between theory and practice remains wide and unbridgeable, because practice remains the responsibility of a local authority operating within an outdated legal and institutional framework.

TYPES OF PLAN

Although several types of plan have been discussed above, it would be useful to review systematically the various possible types of plan. To do this it is necessary to go back again to the fundamental law, no. 1150 of 1942, which gave a more disciplined structure to town planning than had previously existed when planning simply involved the definition of building lines. The law was approved when Italy was at war and just before the dramatic events surrounding the downfall of the fascist regime. The spirit and political purpose of the government which sponsored it were clearly expressed in its objectives and centralised procedures.

It provided for the three types of urban plan which have been discussed earlier. The first of these is the *Piano Territoriale di Coordinamento* covering areas specified by the Ministry of Public Works, not necessarily corresponding to regional, provincial, or commune boundaries. The second is the principal form of plan, the *Piano Regolatore Generale Communale*, which must cover the whole of the territory of a single commune. And the third is the *Piano Regolatore Intercomunale* (PRI) for the whole of the territory of two or more adjoining communes. Although these plans apply at different territorial scales, the PRGC and PRI are essentially plans of the same degree of detail, and in neither case is there a legal requirement that its validity depends on being preceded by a PTC. All three are instruments of urban policy with a programmable content.

There are also three forms of implementation plan. The first is the *Piano Particolareggiato Communale* (PPC), by which sections of a PRGC are implemented by the commune; and the *Piano di Lottizzazione* (PL), by which parts of the PRGC are implemented by any other person or authority. Thirdly, there is the *Piano di Comparto*

Edilizio, which is in effect a scheme of construction or rehabilitation for an area within a PPC. The mayor may invite landlords to develop according to the scheme, and can expropriate property if they do not do so.

Finally, a set of building regulations are adopted by the commune, setting standards to be followed when applying to the mayor for a licence to build. In practice, these may be incorporated within a PRGC.

During the period 1946–55 reconstruction plans were prepared for repair of war damage, and from 1949 special plans were prepared to co-ordinate attempts to solve the housing problem. New residential areas should have been of a guaranteed quality, but with few exceptions they proved to be agglomerations of somewhat squalid and cheap housing. These schemes were promoted by Institutes of Popular Housing which were descended from bodies established in the fascist period, and with greater consistency of quality by a central government agency known as Ina Casa, which had its own architectural and planning staff.

In 1967 legislation known as the Bridge Law, because it was intended to be the first element of a general urban reform, gave the PL type of implementation plan greater legitimacy. It had been intended that the PL would be used only in exceptional cases by private enterprise, but it had proved to be a vehicle for many abuses. These plans were in fact exactly the same type as the PPC, but payment for roads, drains and sewers, and main services in the area of the development, plus a contribution to the expense of providing other public services, was to be made by the private developer the total sum being determined by the local council.

In the 1970s regional reforms were implemented in accordance with the provisions of the constitution, and the fifteen ordinary statute regions were created alongside the five special statute regions already established for Sicily, Sardinia and the frontier areas. The regional authorities have responsibility for planning, and it is envisaged that the regions will develop a strong role in planning, working jointly with the commune authorities. A hierarchy of plans is intended, in contrast to the 1942 system, with a regional plan providing a basis for all decisions within the region, intercommunal and district plans at an intermediate or subregional level, and commune plans. Regions are attempting to achieve stricter and more efficient control over private development. The regional authorities operate the national planning legislation, but also exercise legislative power themselves and may adopt national directives to local circumstances. Regional jurisdiction was defined in a decree in 1977, which also included an interesting definition of urbanism: 'the administrative functions relating to urbanism concern the discipline of the use of land in all its aspects including the operation of safeguards and transformation of land as well as protection of the environment' (DRP616/1977). This extends the scope of urban planning beyond the narrow concept in the 1942 law, implying that urban, physical and environmental planning should be seen as a comprehensive activity. In spite of the innovations in this decree, traditional divisions of responsibility still remain, due to the organisation of government departments established following the constitution of 1948.

In addition to this strengthening of the role of the regions, urban planning procedures were also improved in 1977. Law 10 of that year, which has already been referred to, significantly strengthened the planning powers of the communes by establishing the principle that authority to develop is not vested in ownership rights, but has to be granted by the mayor. Implementation plans and the PRGC were required to be more closely integrated and related to a timescale for development founded on thorough analysis, forecasts and projections, and on a realistic view of what is feasible. The commune administration had to become an active participant in the process of planning, and not simply an agency of control, reacting to planning proposals from others.

A further reform of the late 1970s came in 1978, when a new law concerned with residential development created new procedures and incentives for rehabilitation. Communes acquired the power to prepare plans for rehabilitation, which would identify areas where policies of conservation, reclamation, redevelopment, reconstruction, or improved use of land would be appropriate. Such areas may include single buildings, blocks of houses and areas of service buildings. Both the commune and landlords are responsible for implementation, and their attention is drawn by the law specifically to the need to have regard to existing conditions. This is a new feature, balancing the need to promote development with a concern for environmental conservation and recognition that there is an economic value in conservation. This attitude also reflects that found in the context of ancient sites whose economic value is now recognised after many years of debate about their historic and cultural value.

CONCLUSIONS

One can briefly summarise the main points quite simply. The evolution of planning has led to the commune and regional authorities acquiring prime responsibility for planning, and the principal instruments of planning are the PRGC and the multiyear action programme. Planning the national transport network remains the responsibility of national government. Specific regional planning instruments are being developed by the regional councils, but regional government is still too recent in origin to draw any conclusions about the organisation and management of planning.

Chapter 4
France

P. M. B. Chapuy

The current operation of planning in France must be seen in the context of three major factors. First, France has the greatest land area of any of the EEC nations, with a population of 53 million. The average density is therefore low, at about ninety-seven persons per square kilometre. Over the postwar period rapid population changes have taken place, due to migration to urban areas, high birth rate and immigration. Secondly, France has a specific structure and hierarchy of government, both central and local. Thirdly, the national code of written planning law is mainly administered by agents of the national government.

Responsibility for planning is shared between central government through its regional and subregional or *département* offices, and the lowest level of local government, that of the communes. The responsibilities of these local authorites, though almost identical in law, are in practice very much a function of their size, resources, technical expertise and political will. A variety of specialist agencies has been established by government for specific planning purposes. Co-operation on planning matters between small local authorities is often necessary.

Regional or national planning is mainly economic and structured around the series of four-year plans initiated in the postwar period. They set out the major national, social and physical objectives of public policy, which are then regionalised.

The current system of local physical planning was initiated in 1945 with the building permit which is the basis of the development control system. The present system of plans was established in 1967. The latest major step in planning law was the *réforme de l'urbanisme* of 1976 which brought a more qualitative emphasis to development plans, created other positive planning tools and strengthened the power of public agencies to control development.

One must note the still-important role and power of the political delegate of the central government, the *préfet,* who has power over all the subregional services of the different ministries and is the local representative of each minister. The balance of power between central and local government is fairly equal in the large urban centres or conurbations, where the head of the local authorities, the mayor, is often a political figure of national status (Paris, Marseille, Bordeaux, or Lille, for instance) and where local authorities have large technical services of their own.

The regional or subregional organisation of central government departments is responsible to the *préfet*. A major role in the planning process is played by the Direction Départmentale (Régionale) de l'Equipement (DDE). In addition to these, public or parapublic agencies have been established, including regional research institutes (OREAM) for major conurbations, and the agency responsible to the Minister for the Plan for preparing national and regional economic development strategies (DATAR).

The sharing of responsibilities for planning between central and local government sometimes leads to conflict due to their having opposing objectives. This may explain the presence of large numbers of joint and *ad hoc* agencies and commissions that exist at various levels.

POSTWAR EVOLUTION OF THE PLANNING SYSTEM

France emerged from the Second World War with the enormous task of reconstructing much of its public infrastructure and communications, and modernisation of old housing. Massive projects and redevelopment were instituted. The rapid restoration of the economy led to a very high rate of growth of the towns and conurbations as people moved from the rural areas, and many features of the planning and housing system originated during this period.

NATIONAL ECONOMIC PLANS AND NATIONAL PHYSICAL PLANNING

The overall structure of planning follows a hierarchy from national to local level, and seeks to interrelate physical planning with social and economic planning. The central element is the national plan which sets economic, social and physical planning objectives every four years. All relevant government departments and specific commissions for housing, industry, energy, education, culture, health, transport, and so on, are involved. The plan is approved by Parliament as a programme of public action. Responsibility is allocated to the offices of the regional *préfets* and to the regions, and it provides a framework for regional and local planning.

National directions for physical planning are developed as well. For instance, in the 1960s the national-development pattern for a set of major conurbations or growth poles was established. This concept of *métropoles d'équilibre* was created to orientate growth away from the Paris region and boost the development of the designated areas. The major growth poles of Fos and Dunkirk as well as the Seine valley, thus, included both industrial developments in energy and steel-making sectors and major harbour facilities related to the new pattern of international trade. National policy also includes assistance for poorer regions. For instance, the 'electronic plan' for Brittany was intended to create a new set of interrelated activities in the 'foot-loose' electronic sectors. Policies included industrial subsidies to private firms, the establishment of large research centres of the national Post and Telecommunications Office, the decentralisation of major schools of engineering in the related fields and support for specialised university departments. Special local agencies have also been established to promote tourism development of the Languedoc and Aquitaine coasts.

A national parks system was created in 1960, and five parks now exit. Their purpose is both to protect sparsely populated central areas, and to combine conservation with tourist development. Another innovation in 1960 was the *secteurs sauvegardes* law. This measure protects and enhances areas of towns of historic or architectural interest by means of a specific local plan, stronger control measures and public investment in building or renewal.

DEVELOPMENT PLANS: A TWO-TIER SYSTEM

During the 1960s zoning plans allocating land and defining limits of development were prepared for towns or groups of rural communities. The present two-tiered system of plans was established following a fundamental change in planning legislation which occurred in 1967 with the *loi d'orientation foncière*, or Land Guidelines Act.

At the strategic level there is the *scheme directeur d'aménagement et d'urbanisme* (SDAU), which has to be prepared for urban areas and their close rural hinterland or areas subject to pressure for development. They should express strategic policies for twenty or thirty years ahead in terms of major public investment and the general distribution of urban and rural growth areas included in the plan together with the protection of rural or natural zones.

The second type of plan is a local plan, the *plan d'occupation des sols* (POS), which is mandatory and has to be prepared for all towns over 10,000 people or subject to high pressure (for instance, in coastal or tourist areas). This document consists of a land-use map with an attached set of regulations which precisely define the rights of development attached to the land. The plan is on a base map which enables each individual plot of land to be defined. The POS, after public inquiry and approval from local authorities and the *préfet*, becomes part of the planning regulations. It is then binding on development control decisions over individual or public proposals. Individual proposals are accepted only if they comply with the content of the POS. In areas not covered by a POS development control is based on national regulations, the *règles générales d'urbanisme* (RGU), which may leave more room for discretion to be exercised on individual proposals.

POSITIVE PLANNING TOOLS FOR PUBLIC AGENCIES AND AUTHORITIES

In addition to development plans and control, certain positive planning tools enable public agencies to act as developers or on the land market, and therefore complement or counteract free-market forces. The two main examples are the *zone d'aménagement concerté* (ZAC) and the *zone d'aménagement différé* (ZAD).

ZACS are based on the law of 1967, and have been in operation since 1969. The *préfet* together with a local authority may declare a ZAC, allowing the use of compulsory purchase powers and facilitating the integration of public and private capital. It provides for contractual arrangements between public and private developers for land assembly, infrastructure investment and development in accordance with an ageed comprehensive plan. Many examples have now been implemented, including urban renewal schemes such as La Defense and Front de Seine in Paris, housing projects and industrial or commercial schemes in most major cities.

The ZAD was instituted in 1962 and is a powerful instrument. It defines an area within which a right to acquire land, valid for fourteen years, is given to the state or another public body. About 1 per cent of France is now covered, for purposes such as new towns, protection of rural areas threatened by urban development, especially around Paris, and for tourist development. Land for sale must be offered first to the public authority, at a price based on existing use values.

Since the early 1960s legislation creating wide powers to develop new towns has been enacted. It is partly based on the British principle of completely new and self-sufficient development implemented by an *ad hoc* agency. This agency is controlled by the state and communes affected through the board of the new town. Among the nine new towns so far created, six represent an attempt to redirect the growth of conurbations: Cergy-Pointoise, Evry, Marne-la-Vallée, Melun-Senart and St Quentin-en-Yvelines around Paris, plus Lille-Est in the north. The other three are more on the British model, being self-sufficient towns in prosperous regions. The population of the nine new towns passed from 400,000 in 1968 to 729,000 in 1975 and is expected to have reached 1.1 million by 1982.

QUALITATIVE ASPECTS OF FRENCH PHYSICAL PLANNING

During the 1960s and early 1970s French planning was characterised by a technocratic approach, uniform design solutions and construction programmes that were massive in scale and volume. Changes in ideas, coupled with an end to the period of steady economic growth, led to a change of attitudes during the 1970s. The trend is now towards a more human scale of development, smaller projects and urban rehabilitation, and a greater interest in conservation and respect for architectural history and expression of local culture. Vernacular design is supported by design guides and the adaptation of historic buildings for new uses.

An increasing public desire to be involved in planning decisions is also apparent, and Green Party candidates have achieved some success. Some devolution of planning responsibility from central to local government had taken place before 1981, allowing local expertise both professional and political a greater influence, although central government retained strong financial and legal controls. However, there is not yet any legal requirement for public participation in the early stages of planning procedures. Public inquiries into proposed plans, and some rights for recognised associations to make representations when plans are being prepared, are the limits of the legal provision, but in practice public participation is often more extensive than this implies, particularly if the local politicians are sympathetic.

Another aspect of greater public interest in planning issues is concern for ecological balance. Protection of the natural environment is now a public obligation, following the nature-protection law of 1976. This law requires compulsory environmental impact assessment for large or medium-size projects, both private and public. Analysis of the state of the physical environment and the measures taken to protect or enhance it now must form a specific part of the SDAU and POS. Nevertheless, in both respects practice is still far from satisfactory.

DEVELOPMENT PLANS TODAY: THE SDAU

The two types of development plan referred to above are now well established, although emphasis is now more on the POS. The SDAU plan fixes the basic guidelines for the development of the area concerned and determines, in broad outline, the use to which land is to be put, the layout of major infrastructure, general organisation of transport, location of the most important services and activities, and the areas demarcated for urban extension or renovation. Its purpose is to guide investment programmes of central and local authorities and public services. Public development must be consistent with the plan and express investment programmes for a ten-year period.

The form of the SDAU is a written report and diagrammatic plans. These may include a map of existing land uses and plans showing the situation expected after twenty-five years, and for the intermediate period of ten years ahead. The plan is prepared by a joint commission, which includes representatives of local authorities and the various central government departments. The bulk of the technical work is done by the DDE, although some private consultants are sometimes used. Approval is by the *préfet* after the local authorities have agreed, or for areas of national concern, by the minister.

By late 1980 about 370 SDAU had been started, covering about 26 per cent of the country and 70 per cent of the population. Of these, 164 had been approved, covering 10 per cent of the country and 38 per cent of the population. The average population within each plan area is 120,000. The preparation period and the plan period are both rather long, and revisions are often necessary since plans can easily become somewhat removed from reality. The SDAU have now lost their value in many areas, because the contents have in many cases become out of date, or as a result of political controversy.

DEVELOPMENT PLANS TODAY: THE POS

The more detailed plan, the POS, defines strictly and precisely for each plot of land the uses authorised and the conditions for approval of development regarding size of buildings, form, plot ratio and architectural aspects, and establishes the protection of agricultural and national areas including forests. The POS may provide for the extension of roads and other public services by zoning land to be acquired compulsorily by some public body.

Initiative for its preparation rests with the *préfet* or the local authority. The latter prepares a draft plan, report and regulations, often with the help of private consultants. This is passed on to the DDE, who set up an interdepartmental working group including commune representatives to revise the draft and undertake consultations. The public may be informed, but this is not compulsory. The proposed plan is published by the *préfet*, who holds a public inquiry and the final version has to be approved by the local authority and the *préfet*. The local authority is deeply involved at every stage of the preparation of a POS.

By late 1980 3,427 had been approved, 5,413 published and 11,403 were in course of preparation. The average population covered by each POS is 7,000. Once published, the POS can be used as a basis for development control, for which it is a very precise guide. In theory, all applications to develop within a POS require only a straightforward administrative decision.

CONTROL OF DEVELOPMENT

The major tool of development control is the *permis de construire*, or building permit. It is necessary for any new building, even without foundation works, change of use, change of external aspect, or creation of new floorspace within the building. Other internal works, and repair work by public undertakings such as gas, electricity, or road authorities, are exempt from control.

Formal responsibility rests with the mayor, but most of the more significant decisions are made by the *préfet* and the DDE, with the advice of appropriate government or departmental commissions representing interests of central and local government, and registered organisation.

Development control is often perceived by the public as the responsibility of the state, especially as it is the *préfet* who often signs refusals while the mayor signs approvals. Control is now being brought closer to the public by giving more powers to local authorities which have adequate technical services of their own and an approved POS. This has occurred in seventeen towns.

In most cases outside these seventeen towns consideration of an application is largely handled by the DDE, who is responsible for informing the applicant whether the *préfet* or mayor will issue the decision, and by when. Failure to issue a decision within the period specified constitutes deemed permission. The permit is valid for one year only.

The public is not very well informed, although a public register is kept. Public inquiries are held only for certain environmentally significant proposals. Environmental impact assessment has been compulsory since 1978 for major proposals and operates as part of the control process.

There is no specific planning appeal procedure, although one can appeal to the administrative courts on the grounds that the legal procedures have not been properly followed. This does not include appeal against policy embodied in a plan on which a decision is based.

There are a number of other authorisations of a planning type that apply in certain circumstances, for instance, demolition, opening playgrounds or recreation facilities, felling trees and the control of advertisements. The latter became a local authority responsibility in 1979. One important control is that over the creation of new plots of land. When a developer divides a site into individual plots for development, he is creating new property rights. This has to be authorised by the *préfet*, after a layout plan for the whole development has been accepted. It is, in effect, a form of outline planning permission, since individual buildings would still require permission.

44 Planning in Europe

RESPONSIBILITY FOR PLANNING: CENTRAL GOVERNMENT AND GOVERNMENT AGENCIES

The role of central government, or *l'Etat*, is still paramount. The dominant characteristic is the role of the subregional branches of all major ministries (Environment, Agriculture, Transport, Health and Social Services, Education and the Interior). These technical branches come under the authority of the *préfet*, a high civil servant named by the Cabinet and who is involved in all local matters. The DDE, a branch of the Ministry of the Environment, is mainly staffed by engineers. Only since the 1970s has a training in planning become a regular part of their education. Graduates in geography, economics and law are now to be found in the DDE as well. The DDE often takes the lead in preparing development plans and offers technical advice to local authorities. The *préfet* has the power to decide most local planning issues, and is personally involved on important cases or when the DDE and local mayor disagree.

There are a number of other government agencies of interest. Strategic studies and policies for economic development are prepared by a central agency known as DATAR. There are also ten research and advisory bodies, responsible to the regional *préfets*, for advising on economic and physical planning known as OREAM. Agencies to supplement the local DDE have been established in coastal areas to promote the planning of tourist-related development. New towns are designed and implemented by corporations with their own technical staff, which have exclusive planning powers within their designated areas, although local authorities have some control over them. National parks also have their own technical staff, responsible to a board on which central and local government, academics and local interests are represented. They are financed by central government and have exclusive responsibility over the central core of the parks (the protected natural and wild areas) and consultative power over the surrounding area, in which tourist accommodation and facilities might be located. Other special agencies and commissions are established from time to time for specific purposes. One example is a governmental agency responsible for defining areas suitable as ski resorts, and controlling large development schemes.

RESPONSIBILITY FOR PLANNING: LOCAL AUTHORITIES

There are three types of local authority. At the most local level are the 36,000 communes. Above these come the ninety-five *départements*, which have the same territory as the subregional *préfets* and branches of national ministries; and the twenty-two regions. Up to 1982 these two types of authority had mostly investment powers on specific programmes such as transport and social or health services, as well as economic planning tasks.

The communes were created in 1789, with populations ranging from ten inhabitants to 2.7 million in the case of Paris. A council is elected, and committees for specific tasks are formed. The mayor is chosen by the council, and he is both political leader and head of technical services. Some employ no staff, and some employ several thousand. Planning work is shared with the DDE, and the extent of the commune's own contribution depends on its size, staff and resources, and the political stature of the mayor and the council.

The modern trend is to allow the communes more independence, once they have the expertise and an agreed POS or other plan. Communes may group together to co-ordinate their responsibilities for sewage, waste disposal, housing, planning, or transport, for example, by creating a special joint authority. There are also twenty-eight *agences d'urbanisme*, which are advisory planning agencies for groups of communes, financed by central and local funds, responsible for preparing planning studies and policies for the whole of an urbanised area.

Planning decision-making in local authorities is in the hands of elected local politicians. Local councillors do not receive a salary or work full-time for the commune, although the mayor of a large authority may receive sufficient financial allowances to enable him to become almost a full-time executive.

THE PLANNERS

It is not possible to identify planners as members of any one professional institute, or the holders of a specific qualification. The majority of planners are employed by the DDE, large communes and firms of consultants. About 2,000 professional planners work in the first and second of these, that is, the public sector, of whom about half are preparing plans and half controlling development; and in the private sector there are about 1,000 professional planners. The overwhelming dominance of engineers is declining now, as more geographers, social scientists, or engineers with planning qualifications are employed in the public sector. Most communes have very few technical staff, with no professional planner, but the number of planners in local government and local agencies is growing. Private firms tend to employ mainly architects and architect-planners. This sector is no longer growing, due to the decrease of building activity and the growth of the public sector.

Planning education has evolved rapidly from the situation in the 1960s when it largely took the form of urban-design courses for architecture students. These courses still operate, but two other main sectors of planning education are now most important. These are in the universities which have specialist planning courses, often at a postgraduate level, and the engineering and administration *grandes écoles*, where students may specialise in the final year in planning. Almost all senior civil servants graduate from these schools. The postgraduate students come usually from a background of engineering or architecture and to a lesser degree from geography, sociology, or law. Besides these professional courses, there is a strong tradition in many universities of research in topics related to modern urban society from disciplines such as sociology, politics, or philosophy, often based on Marxist or other political ideologies.

The lack of a professional qualifying institute or protected title for planners means that their status is often lower, or less secure, than that of engineers, architects, or other professions, especially in government service. The association that does exist, the Société Française des Urbanistes, has not yet acquired a very high status.

LEGAL AND CONSTITUTIONAL BASIS

The national government initiates legislation, which has to be approved by Parliament. It also issues regulations and circulars which are strictly applied by the appropriate branches of ministries. Planning law is entirely a written code of law, embodying everything from the national *code de l'urbanisme* to local zoning and regulations. Central government can issue regulations on national policies as it does, for instance, in mountain and coastal areas. These regulations have the same legal status as the RGU or the POS and they must be complied with when deciding on development proposals. In addition, the contents of SDAU and POS must be consistent with them.

French planning legislation includes provisions by which public and private actors and developers jointly may undertake positive planning. These include land assembly, public infrastructure development, housing, offices, or industrial development.

The major legislative provision since the war may be summarised as follows. In the 1940s the building-permit system and the first type of development plans were established, and the national socio-economic plan came about. The 1960s saw the National Parks Act (1960), the Protected Historic Areas Act (1960), planning acts in 1962 introducing the ZAD and the Land Guidelines Act (1967), the latter creating the development-plan system of SDAU and POS, plus the ZAC. In the 1970s came the Regional Reform Act (1972), the Land Reform Act (1975) and the Nature Protection Act (1976), which introduced environmental impact assessment and led to environmental concern in physical plans. In 1976 the Urban Reform Act, which included

France   47

provision for public participation and strengthening of the RGU, was also passed.

CONCLUSION

Central government, through its ninety-five branches, and local authorities are involved in daily partnership in physical planning, in its plan-making, development control and positive planning aspects. The structure of public administration gives the system a high capacity to coordinate and implement action at the regional or subregional level, involving several local authorities, and in pursuit of nationally defined policies. Policies can be immediately applied and enforced by the ninety-five administrations throughout the country.

The powers of the communes in planning are very much related to their size, technical expertise, financial capacity and political will. Specific joint bodies grouping several of them can be set up to develop these capacities. As their size varies greatly, formal responsibilities as well as resources may vary.

The development plan system has two tiers of physical plans, related to a national socioeconomic plan, the latter being regionalised. Development control is conceived primarily as an administrative task based on precise regulations, in the form of either the national set of regulations (RGU) or the approved local plan (the POS). The content of the latter is binding. The decision is often the legal responsibility of the *préfet*, especially for large developments. Any appeal may be made through the usual administrative court, but for maladministration only, that is, only the procedure and not the content of the decision can be challenged.

Public participation is legally required only at plan-approval level or for the evaluation of specific developments. Registered *associations* may be heard at plan-preparation level and have the right to appeal against any administrative decisions if they think appropriate.

Protection of the natural environment is ensured by specific protection measures of different strengths, of which the three major categories are national parks, nature reserves and protection of sites of landscape value. Also public and private projects may be subject to environmental impact assessment and all public plans and policies, including development plans, must now take account of environmental issues. This practice is not yet strongly established, however.

Positive planning is encouraged by means of various devices to facilitate joint working between public and private sectors to develop land together.

Professional planners have up to now been mainly engineers in central government, and architects in private practice. The situation is changing rapidly, with more planning education for engineers, growing local authority planning departments and the inclusion of social scientists in planning teams. Planning is not, however, recognised as a profession in its own right.

These changes, and changes in the qualitative concerns of planning for the environment and a humane scale of development, are reflected in the successive titles of national ministries responsible for physical planning; Ministère de la Réconstruction, Ministère de l'Equipement et du Logement, and between 1978 and 1981 Ministère de l'Environnement et du Cadre de Vie.

Following the election of President Mitterrand, and the change of majority in the first chamber of Parliament in 1981, a wide-ranging programme of decentralisation is being put through. The recent Decentralisation Act 1982 is going to alter greatly the present balance of powers, duties and responsibilities. The outcome is likely to be a fundamental extension of the powers and responsibilities of the communes, *départements* and regions in town planning for plan preparation, positive planning and development control, and to a lesser extent in economic planning. At the time of writing it is still too early to describe the effects of these changes.

Chapter 5
Netherlands

H. van der Cammen

The main circumstances determining the character of Dutch planning are the high density of population and pressure on space, the interplay of conflicting interests and the delicate network of decision-making responsibilities. The legal framework has been more or less stable since 1965, but was in recent years supplemented and supported by new judicial and financial instruments. Planning practice has shown progressive refinement in its style and methodology, reinforced by the development of specialist skills and their mutual integration in the planning profession. In the near future a narrower definition of the scope of planning is likely to be adopted, together with correspondingly stricter professional requirements.

As in other countries, national and regional planning grew from experience with planning on a local level, while the latter developed from housing and building regulations. The first land-use plans (town-extension plans, regional plans) were developed during the 1930s. Prewar evolution of ideas and concepts was particularly influenced by British leaders of opinion such as Abercrombie, Geddes and Unwin. In the meantime, a group of young Dutch architects and designers (for instance, van Eesteren) played a prominent role on the continent in the CIAM (International Congresses of Modern Architecture) movement. This connection is easily recognisable in the general development plan for Amsterdam of 1935, which together with the general plan for the reconstruction of Moscow in the same year, marked the start of long-term town-development planning in continental Europe.

A second very unusual characteristic of prewar Dutch planning is its constant striving for institutionalisation at the national level, which led to the establishment of a National Physical Planning Agency (Rijksplanologischedienst) in 1941. The ambition of adopting national physical plans has never been realised; instead a practice of issuing periodical policy statements on general topics connected with spatial development has grown in postwar years.

The national planning policy during 1945–65 reflected the national objectives of general economic recovery and improvement of the housing stock. In the first National Statement of Physical Planning,[1] in 1960, the position of the western regions (Randstad Holland) as the driving-force of the economy was consolidated, while at the same time policies to achieve a more even distribution of the population, and economic growth in the periphery, were proposed. Regional plans were devised during this period with the same objectives. However, apart from strong demographic and economic growth in Randstad Holland, these policies did not succeed.

At a local level many town-extension plans (*uitbreidingsplannen*) accommodated the rapid postwar expansion of population in ways which were efficient in a technical sense. As a consequence of the loss of housing stock during the war, and the unexpected growth in *per capita* living space, this population growth created an immense demand for housing. The years after 1965 saw a rapid decay in the quality of the older existing housing stock, and at the same time a wide range of new entrants into the housing market, with the result that the housing problem in the Netherlands is as urgent now as it was twenty years ago.

In 1965 the Physical Planning Act came into force, and in 1966 a second National Statement on Physical Planning was issued.[2] These events mark the beginning of a new period in Dutch planning practice. The central problems which have faced planners during this period are the urbanisation of the countryside, especially manifesting itself in the growth of many small villages, a traffic boom and deterioration of the environment. The response from the government in the second National Statement was a design or blueprint for 'collected suburbanisation' for the year 2000, a policy to divert sprawl into regional growth poles combined with a renewed effort to stimulate the north-eastern and south-eastern periphery. In the years that followed it became clear that in the absence of adequate policy instruments this new concept would remain ineffectual. Eventually this reflected the lack of political will to put the concept into effect.

The 1965 legislation created a substantially new planning framework. New-style regional plans were devised to guarantee co-ordination of public policies at the different levels. A machinery for co-ordination was established in the form of standing committees of professional representatives from various agencies at national and regional level. Only from 1975 onwards have the instruments necessary to achieve the old concept of growth poles come into operation. The third National Statement on Physical Planning was issued in 1976,[3] and the urbanisation section finally dropped the longstanding but largely unsuccessful policy of economic and demographic stimulation of the peripheral regions and replaced it with a policy of conservation. Containment of growth in the Randstad was complicated by the considerable scale of urban renewal, which forces central government to pursue a policy of strategic land-use allocation for city regions.

Local planning since 1965 has been increasingly dominated by the complexity of planning procedures. The *bestemmingsplan* (literally, 'destination plan') created by the Physical Planning Act 1965 can serve as a tool for both conservation and development. The Dutch administrative system, in which the planning system has to fit, is characterised by both a long tradition of central rule and a strong emphasis on the concept of legal certainty (Rechtstaat). Both features are easily recognisable in the administration of urban and regional planning, especially in the *bestemmingsplan*, and this leads to abuses on both sides. Public authorities are inventive in finding loopholes in the inflexible system of law, and appeals by individual citizens can postpone the implementation of development plans, sometimes for several years. The threat of urban congestion led several local authorities in the late 1960s to undertake large-scale redevelopment of city centres, but during the 1970s this trend has been replaced by small-scale rehabilitation.

Economic prosperity in the 1960s led to the general acceptance of economic, physical and sociocultural planning as three equivalent facets of government planning, providing overall direction for a number of smaller-scale sectoral planning processes. The economic crisis of the late 1970s, however, highlights weaknesses in the basic co-ordinating role of physical planning. Sectoral agencies responsible for traffic, agriculture, public works, and so on, exercise the influence of their large budgets to implement infrastructure and housing projects, while physical planning authorities generally lack sufficient control over investment to achieve effective co-ordination.

A final problem is the constitutional position of public authorities with respect to private organisations, industries, developers and even citizens' groups, which has given rise to a Dutch version of the permissive planning model. Faced with private capital and private influence Dutch planning authorities seem to lack power, resources and, indeed, courage to give a lead in the processes of physical development. The judicial system of building control has frequently proved insufficient for the purposes of positive planning. Consequently, the quality of physical development in the Netherlands is to a significant extent the result of decisions taken in the private sector, with public authorities responding to the trends of the time from a certain distance.

PLANNING PROBLEMS ON THE GROUND

The pattern of settlement in the Netherlands is characterised by the presence of many small but concentrated urban settlements. Attention in Dutch national planning has been divided between urban problems and the problems of the countryside. Two national policy reports have been published recently as part of the third National Statement; the Urbanisation report (1976) and the Rural Areas report (1977).[4] The complementary fields of national planning constitute sectoral and infrastructural developments. In 1979 the number of inhabitants of the Netherlands exceeded 14 million. Of this number, around 50 per cent live in cities with more than 50,000 inhabitants. In 1978 the steady reduction in the Dutch birth rate slowed down. Foreign immigration will remain substantial but the population size is expected to attain a maximum of around 15 million by the end of the century, after which a gradual decline is anticipated.

The most urgent and persistent planning problem in Dutch urban areas is the constant housing shortage. Demand for housing recently received a new impetus from urban renewal, by a considerable increase in divorce rate, and by the formal acknowledgement of independent housing needs of all persons of 18 years and older. Supply is complicated by the rising costs of building and a comparatively low level of acceptable housing expenditure. The housing problem is serious, especially in the big cities in the western part of Holland (Randstad). Urban overspill measures concentrate on growth centres and growth towns. About a dozen small municipalities, designated as growth centres in an outer circle around the Randstad, have been commissioned as new locations for housing projects. Some 35 per cent of all new dwellings are in this region. However, this national policy of diverting growth in the direction of new centres at a distance of 30–50 kilometres from the central city is under attack. New centres receive special financial support for expansion, but problems of organisation and co-ordination cause considerable delays in implementation. In the meantime there is a bottleneck in the resettlement of overspill population from renewal areas within the central cities.

Largely the same problems are experienced in so-called growth towns, five larger cities designated to accommodate large housing projects in the northern, central and southern parts of the country: Groningen, Zwolle, Apeldoorn, Breda and Helmond. The substantially increased mobility, coupled with a high density of population (380 persons/km^2) causes exceptional problems. In spite of rising energy costs, there is constant pressure on the already very dense network of roads. The demand for public transport is also displaying a slow but constant increase, which has to be met by considerable new investment. In national policies designed to curb the growth of mobility major emphasis is given to the concept of a city-region: a central city in which services, particularly employment, are located, and a surrounding area which is functionally oriented towards the centre.

Lack of amenities for residents in many small villages and increasing conflicts in the use of open areas are major problems of rural planning.

In some 15 per cent of villages of up to 1,000 inhabitants there is at most one shop for daily food. Facilities for primary education and for social-cultural activities are diminishing, even in villages with more than 1,000 inhabitants. This deterioration of the level of services has varied effects on different population categories and provokes the growth of mobility in the countryside. Local authorities try to meet the situation by creating dispersed housing projects which brings them into conflict with the concentration policy of the national government. In rural areas there is conflict between three activities: agriculture, recreation and conservation. In the urbanised western part of the country agricultural areas of high value are sacrificed for recreational use of the adjoining metropolitan area. On the other hand, the rationalisation of agricultural production is threatening the quality of landscape and environment. A policy to implement national and regional parks has started recently. Environmental protection is increasingly identified as a long-range policy issue.

The combination of high density of population and physical conditions poses specific questions of infrastructure: the location of a possible second national airport (actually planned on a proposed Ijsselmeerpolder which still has to be reclaimed), the location of new power stations with the associated questions of the use of nuclear energy and storage of nuclear waste, and sites for extraction of minerals (sand and gravel). Recently possible conflict in the use of the Dutch part of the North Sea (waste storage, mining, industrial islands and recreation) has received more attention.

Regional planning problems display a large degree of differentiation. The eastern part of the country (North-Holland, South-Holland and the western part of the province of Utrecht) was recognised as a congestion area back in the 1950s. Now it is called a problem area. This region experiences considerable traffic congestion, problems caused by concentrations of industry and a steady increase in average demand for space per inhabitant in an area of rather high density (900 persons/km^2). The level of services is deteriorating, especially in the central cities, while unemployment is above average. Migration balances are negative (13,000 to 27,000 persons a year in the 1970s).

The northern (Groningen, Friesland and Drente) and the extreme southern (Limburg) provinces also constitute problem areas for social and economic reasons. For both areas, regional policies of economic restructuring have been developed. Relocation of government services from The Hague to Groningen and Limburg is considered a main element of policy. The postwar migration from the north to the Randstad in search of employment has recently stopped, probably in reaction to the growing congestion and housing shortage in the west. In the extreme south unemployment started in the 1960s with the closure of coalmines but a negative migration balance developed only recently.

The central parts of the country (Utrecht, the Ijsselmeerpolders, Noord-Brabant, and the western parts of Overijssel and Gelderland) experience increasing locational advantages. Most of the growth towns are situated in this region, which has had a constant positive migration balanced in the 1970s. There is a considerable movement of industries and trade from the Randstad into this 'halfway zone' of central Holland. The region contains a large area of open space, which according to national policy statements has to be strongly safeguarded against urban sprawl. The southern Ijsselmeerpolders (reclaimed in 1968) constitute a new countryside in which two new cities (Lelystad and Almere) are being developed.

Finally, unique subregions are situated in the extreme north-western part (Waddenzee) and the extreme south-western part (Zeeland) of the country. In the Wadden area planning policy is directed towards conservation of the natural environment. Zeeland includes the Rhine-Maas-Scheldte delta in which implementation of the Delta plan has been underway since 1958. This is a system of dams, locks and communications, which has stimulated changes in the functional structure of the region. Previously isolated, the province of Zeeland has experienced increased urbanisation and seen many new recreational developments, due to the drastic reduction in travel time to the major cities following the construction of new dams, bridges and roads.

Planning problems at the local level are dominated by the critical financial situation of most of the Dutch municipalities. This becomes most obvious in the growing stagnation of urban renewal in medium and large cities and in a slow but noticeable deterioration of the quality of life. In the Netherlands the municipality plays a key role in the implementation of housing and public services, but financing has to be assured by means of long and complicated negotiations with departments on a national level. These problems are especially acute in the four big cities in the western part of the country: Amsterdam, Rotterdam, The Hague and Utrecht. These cities show considerable loss of population, an increasingly unbalanced population structure and decreasing employment opportunities. Foreign immigration puts great pressure on the older and cheapest parts of the existing housing stock dating from the late nineteenth century, where in some cases 50 per cent of the population consists of ethnic minorities. The high birth rate of these groups will establish a real multiracial society within the boundaries of the big cities. The local authorities give central consideration to containment of the population losses and recovery of employment. Development of new housing projects in many open spaces within the city area is actually seen as an important planning instrument, but implementation turns out to be an expensive and complicated affair and is not always welcomed by the existing population.

Many cities have recently witnessed squatting, or *kraken*, a spontaneous movement by young people who take possession of unoccupied offices, expensive apartment blocks, residences, and so on. Sometimes these actions result in temporary housing contracts for the new dwellers, but in other cases a major confrontation with the authorities is inevitable. City planning in the Netherlands is characterised by a strong involvement of the population. Many *ad hoc* and permanent action groups work zealously to influence decision-making, and many organisations and pressure groups are deliberately drawn into plan-making by the government.

PLANNING AGENCIES

Plan preparation is commonly dominated by government agencies, but strategic decision-making and implementation are increasingly influenced by the presence of private organisations. On a local level planning responsibility is assigned to the elected municipal councils. In practice most of the daily planning powers are concentrated in the hands of the executive board of *burgemeester* and *wethouders* (mayor and aldermen), the first appointed by the Crown and the others chosen by and from the municipal council. For the preparation and implementation of planning policies local politicians have two kinds of adminstrative agencies at their disposal, professional, technical and scientific service agencies, and secretariats. The last are devised for all kinds of legal assistance, for the consultation of other agencies involved and for general co-ordination of processes of planning and implementation. Technical and scientific agencies provide the expertise for substantial research, plan-making and development control. Instead of maintaining a permanent technical and scientific service of their own, smaller municipalities commission private planning consultants to do their planning work.

The organisation of planning by provincial governments is similar to the local situation. General policy plans will be adopted in the near future to establish intended policy for a four-year period both in planning and in other fields of provincial jurisdiction with the clear intention to prevent conflict in the relations with local and national policies.

The National Physical Planning Agency, established in 1941, is a part of the Ministry of Housing Physical Planning. Initially its task was to prepare a national plan. However, no national plan, as such, ever saw the light of day. The Agency instead became an important centre of co-ordination of policies at a national level. This turned out to be particularly necessary during the 1970s when physical planning, in the strict sense, became linked up with the activities of other ministries responsible for traffic and transport, agriculture, environment, recreation and welfare. In co-operation with the Physical Planning Agency they formulated planning policies of their own. The position of the Minister for Physical Planning as a central focus of policy development and co-ordination on a national level is made difficult by his comparatively low budget. Funds for infrastructure projects, rural development, and so on, are mostly allocated at the discretion of other national agencies and ministries. For instance, the planning and urbanisation of the new Zuiderzeepolders is implemented under the authority of the Ministry for Transport and Waterways (Rijkswaterstaat). At both national and local level governmental bodies are under various pressures from private organisations. In some special cases this pressure appears to evolve into extended participation in planning and implementation. Interest groups in the fields of industry, agriculture, environmental protection and transport are formally consulted during the process of important national physical planning decisions, and in many other cases their opinions and policies are anticipated and taken into account by governmental agencies. In local planning there is a strong involvement of local interests, developers and action groups. Developers are sometimes entrusted with aspects of plan preparation and implementation of inner city redevelopment projects. There are also several instances of successful intervention of action groups in the development process. Action plans in the cities are sometimes prepared by project groups in which citizens participate side by side with planning officers. Sometimes when municipal planning departments are overloaded, plans drawn up by private bodies are adopted as official policy with only minor modifications.

THE LEGAL AND ADMINISTRATIVE FRAMEWORK

The general framework for public planning is laid down in the Physical Planning Act 1965. The latter makes references to governmental planning agencies, different categories of plans, the legal powers of planning authorities with and without a plan, and regulations for compensation to citizens injuriously affected by public planning measures. Dutch planning law reflects the concept of *rechtstaat*, which is a fundamental principle of public law in the Netherlands. The state is supposed to guarantee legal certainty for the citizens, and the law presents many cases in which this prime public responsibility becomes manifest. An important example constitutes the *bestemmingsplan*, or designation plan, generally considered the cornerstone of Dutch planning law. The Physical Planning Act requires local authorities to prepare *bestemmingsplannen* for new development and conservation in rural areas. Making a *bestemmingsplan* for parts of the built-up area is possible, and in cases of extensive redevelopment, obligatory. The spirit of the *bestemmingsplan* is negative; it is primarily an instrument to prevent undesirable forms of redevelopment in the countryside and in situations of urban renewal. A *bestemmingsplan* is also used for the layout of new urban development and of redevelopment areas. It consists of a plan, supplementary regulations and a separate explanatory note. Plans and regulations concentrate on the allocations of land use in the planned area, the kinds of building permitted, density and height of the buildings. It is prepared by the municipal council and needs the approval of the provincial authorities. An approved plan is binding upon both local authorities and citizens. It plays a formal role in local development control; development can only be (and must be) permitted if it is in accordance with the *bestemmingsplan*. A local authority that wishes to depart from its own plan has formally to start a public procedure to revise the plan. Article 19 of the Physical Planning Act contains a very popular loophole, by which local governments can by-pass this whole complicated legal procedure in order to achieve more flexible planning, but this works to the detriment of the individual citizen's legal certainty.

Municipal authorities can devise structure plans for the whole or considerable parts of their territory. These plans have a more verbal character; they are the expression of the municipality's general intentions and indicate the location of large structural elements. Structure plans generally have a strategic character and are not subject to approval by higher authorities.

Provincial authorities prepare *streekplannen* (provincial structure plans) for all or part of their territory. Presently around thirty-five such plans, some of them still under preparation, cover the whole country. Provincial structure plans are of ever-growing importance in the co-ordination of physical planning at all levels. The provincial authorities are on the one hand involved in the preparation and implementation of national policy, and on the other supposed to assess the relationship of local *bestemmingsplannen* to the relevant regional plan. Recent plans are also mainly strategic in character. A distinction is drawn between essential and variable elements (the latter can be modified without a formal revision) and different timescales are fixed. By law provincial structure plans have to be revised every ten years, but there is a growing practice of thorough evaluation every five years. The *streekplan* can be generally characterised as a policy programme, containing a broad sketch of the goals of regional policy, administrative and financial instruments for implementation and a design of structural physical elements.

National physical planning is accommodated by a system of structure schemes and structure outlines. The first category contains infrastructural programmes for water supply, energy distribution, major housing location, airports, recreation facilities, environmental conservation, regional and national parks. Integration of all these policy sectors is supposed to take place in two national structure outlines: the *Outline for Urbanisation* and the *Outline for the Rural Areas*. The national schemes and the two national outlines are related like warp and woof. While in the schemes the emphasis is on different categories of infrastructure projects, the two outlines reflect an integrated view of physical development in urban and rural areas respectively. These schemes are prepared after intensive deliberation by the ministries and national agencies concerned. There is also a standard procedure for participation of interest groups and individual citizens after which Parliament is consulted. The procedure ends with a formal policy statement by the government. This procedure, known as core-planning decisions, is still experimental, as is the system of outlines and schemes. As yet, neither is embodied in the Physical Planning Act.

The Act contains a strong centralising tendency. Both national government (that is, the Minister for Housing and Physical Planning) and provincial government (that is, the provincial executive board) have the power to give finding instructions to local authorities to change their plans.

This power to intervene directly in the discretionary field of the lower authority is not used very frequently, but the mere possibility of intervention acts as a restraint on the one hand, and provides an instrument of power on the other. Another special characteristic of the Act is the possibility of individual citizens' upholding their objections to a plan all the way to the Crown. This sometimes causes considerable delays in making a plan operative. For some *bestemmingsplannen*, this period can be up to five years after their adoption by the municipal council. Many other laws are relevant to the public management of physical development. The Housing Act is the base for municipal building regulations. The Act regulates compulsory purchase of land and building in the interest of housing, reconstruction and conservation. Various other laws regulate public intervention in the field of traffic, environment, redevelopment of rural areas, or location of industries. An Urban Renewal Act has been under preparation for nearly ten years.

Alongside the various types of plans under the Physical Planning Act several other kinds of plans are in use. In urban renewal and in new building areas a *bouwplan* (building action plan) is frequently used, not least to avoid or to supplement the inflexible instrument of the *bestemmingsplan*. Rural reconstruction is based on redevelopment plans; traffic planning on a variety of road plans; and the Monument Act allows for the designation of protected parts of towns or villages. With urban renewal, investment programmes serve as important policy instruments. Finally, public authorities make increasing use of reports to state new planning policies without endowing them with the permanence and inflexibility of a physical plan. This practice is especially common on the national level (the Urbanisation and Rural

Areas reports), and on the local level, where municipal councils frequently avoid the long road of preparing a structure plan by formulating middle-range planning policies for the duration of their term of office.

STYLE OF PLANNING

The actual style of planning in the Netherlands is a mixture of strategic and negotiative elements. It can be seen as the outcome of an evolutionary process in the field of planning methodology which started in the postwar years and received special impetus from 1965 onwards. In the period 1945–65 the Dutch planning system developed together with postwar construction. The high rate of population growth, together with economic recovery, created an enormous demand for land-use and development plans. The technique of preparing these plans was rather simple. Social researchers and economists were called upon to prepare projections on future land-use demands, population growth, density, investment capabilities and spending power. Then a design was prepared for redevelopment of an open area into urban space. Usually these designs were made by architects who in later years had a training in techniques of urban and regional design. These designs contained only casual references to preceding social and economic research into the area for which the plan was prepared. Finally, the design was cast in a legal form which made the whole suitable for political approval by the use of proper procedures. Research and design were undertaken by technical and scientific agencies, and the statutory finishing-touch came from the administrative secretariats.

This standard working method produced a long sequence of blueprint plans, which clearly reflected the existing climate of opinion. Conditions for planning changed by the end of the 1960s. Planning methodology entered a process of fermentation which still prevails. The origins of the breakup of traditional methodology are manifold. First of all, the policy environment of planning showed some important new characteristics: political attitudes became relevant to many aspects of policy which were previously considered to be largely technical matters. Urban and regional planning is one of these areas, which from 1965 onwards even became one of the main new areas for concern of politicians and the public. This brought policy preparation and policy formation, previously seen as the business of planners and policy advisers, into the public eye. Another trend, closely connected with the first, is the multiplication of social dimensions of planning. While planning was previously concerned mainly with housing and technical infrastructure, now it broadened into a concern for the whole spectrum of social issues. This brought many new actors into the planning field, and at the same time devalued the practice of blueprint planning. Finally, a third important trend connected with the above was the gradual penetration of the social sciences into planning.

In the beginning these rather unexpected developments resulted in a methodological device known as process planning. In fact, this device was close to methodological anarchy, as it was developed largely from negative reactions to former practice. Long-term planning was scorned, the rigidity of plan preparation was questioned and there was a strong emphasis on continuous and cyclical planning. Generally, however, process planning was limited to declarations of principle until around 1970 the methodology of goal formulation and programme evaluation became popular. Many new plans from this period contain a reasoned structure of goals, disaggregation of goals into concrete policies and methodological indications for measuring the success or failure of the alternatives in the light of the original general goals. These new planning procedures corresponded to the political need for medium-term policy horizons, and to the growing necessity of negotiations with a multitude of actors during the plan-preparation phase. At the same time, a practical answer was offered to the theoretical claims for continuous and cyclical planning. The framework of goal formulation and evaluation allowed the professional planners to adopt a broad range of methods and techniques, developed within social and economic disciplines and learned from foreign literature. The old model of planning (research, design and regulations) was superseded in the 1970s by the project group, involving a horizontal restructuring of plan preparation and policy formulation, easily accessible both for professionals from different disciplines and for active politicians.

The most recent developments show two main characteristics. The original strict procedure of goal formulation and evaluation is making room for a more strategic approach. In this method there is no rigid inference of concrete policies 'top-down' from general goals. General goals are now of diminishing importance. Instead, medium-term policy goals are being formulated in close relation to concrete expressions of desired planning solutions. There is a very pragmatic element in this recent development, which is sometimes clearly expressed by the formal intention to review policy goals every four to five years. Monitoring is also acquiring growing importance.

The second characteristic of current methodology is the continuing dominance of negotiations as a tool of planning. The growing number of actors involved, smaller economic and political margins, a traditional national predilection for debate and the circumstance of a small country with all centres of decision-making within one day's travel are all factors facilitating a climate of continuing negotiations around main planning issues. Increasingly a shift from processing planning solutions into processing continuing planning problems is noticeabe.

These changes in planning methodology have been accompanied by shifts in underlying professional ideologies. While in the 1950s and 1960s, mainly in social science circles but also among some groups of designers, fairly critical attitudes prevailed towards the content of planning policy, considerable changes took place after the growing involvement of practitioners in process planning. The new methodology is characterised by a basic acceptance of the need for short-term change, especially for urgent problems like urban renewal and environmental protection. The element of critical distance which was so characteristic of the methodological position of the profession until 1965, is now replaced for all intents and purposes by the pursuit of effectiveness.

THE PLANNING PROFESSION

The planning profession in the Netherlands exhibits no formal patterns of affiliation. Work takes place in an open context of communication with members of various social, technical and economic disciplines. In association with the evolution in methodology, the planning profession changed from moderate monodisciplinarity and became fully multi-disciplinary. After the initial dominance of architects, civil engineers and geographers, the 1960s saw a gradual widening of the number of disciplines represented, especially from the social sciences: policy and administrative sciences, sociology, psychology and educational science. The input was twofold. On the one hand it reflected the differentiation in the policy content, on the other hand several contributions were made to planning methodology as specialists on policy analysis, cybernetics, evaluation techniques and public participation entered the field.

Sectoral specialisation has made great progress. Most Dutch planners are specialised in one of several aspects of planning such as traffic, public housing, agriculture, natural environment, and so on. The internal structure of the main planning agencies follows the same lines, which is increasingly complemented by project groups in which different professionals are co-opted according to the relevance of their specialism. Modern Dutch literature on policy and methodological aspects of urban and regional planning plays a considerable role as a communication vehicle between members of the profession. There is also a platform for study and contact in the Netherlands Institute for Planning and Housing (NIROV). The institute, which was set up in 1917, is a meeting-ground both for members of the profession and many political representatives. It acts as an influential interest group in questions of public policy and organises study groups for special categories of practitioners.

Planning education is informally related to the differentiation in professional orientation. Two kinds of academic courses can be distinguished. The first category consists of general, comprehensive courses, originating in both technical and social studies, with an explicit planning emphasis. The core consists of planning theory, organisational knowledge and planning techniques, with some elementary specialisation in design, infrastructure, or social aspects. The second category consists of a broad spectrum of opportunities for specialisation within traditional disciplinary courses in urban and regional planning, like urban and regional economics, housing, traffic and transport. Both kinds of educational course produce practitioners of roughly equal status. This structural differentiation in academic education is reflected in the openness of the Dutch planning profession.

CONCLUSIONS

Urban and regional planning in the Netherlands displays a number of special characteristics. The high density of population, in combination with a marked degree of economic and social welfare, gives rise to particularly pressing demands for space and the relatively high political significance of planning. Involvement of public authorities in development and management of the physical environment is traditionally extensive, partly due to the special difficulties of developing and redeveloping infrastructure. The continuing problems of the housing market create a special link between planning and housing policy which is reinforced by the weakness of public authorities in the sphere of industrial location. There is a long tradition of planning especially in the context of zoning within the urban areas, protection of the open countryside and national planning policy. The latter element is traditionally strong, although there has been a trend towards decentralisation to the regional, strategic level in recent years. Well-organised private interests utilise several channels of participation in plan preparation and implementation. Planning methodology is well developed, mostly in the direction of strategic middle-range plan preparation and negotiative capabilities. The planning profession consists of an open grouping of specialists from different disciplinary specialisations with a common expertise in administrative knowledge and with common reticulist skills.

NOTES: CHAPTER 5

1 *First Report on Physical Planning*, The Hague, Ministry of Housing and Physical Planning, 1960.
2 *Second Report on Physical Planning*, The Hague, Ministry of Housing and Physical Planning, 1966.
3 *Third Report on Physical Planning*, The Hague, Ministry of Housing and Physical Planning, 1976.
4 These were published as Pt 2 ('Urbanisation') and Pt 3 ('Rural Areas') of the *Third Report on Physical Planning*, ibid.

Chapter 6
Belgium

M. Anselin

INTRODUCTION

The basis of planning in Belgium is the definitive Law for Town and Country Planning, voted on 29 March 1962. This was subsequently amended in the light of experience by the laws of 22 April 1970 and 22 December 1970. In principle, planning takes place at the following levels: national (although there is no national plan); regional, by both regional and subregional plans; and local, by means of general plans for a whole municipality and special plans for part of a municipality.

The essential mechanism of planning has been by means of permits for land-parcelling, and by building permits. A limited number of municipal authorities have made special plans for part of their areas, for purposes such as zones for industrial development, play and recreation areas, or social housing projects. Only about 200 of the former 2,700 municipalities have so far prepared a general plan.

In the period 1956–6, the central government commissioned a number of regional studies for twenty-five macroregions. The outcome of these studies, known as 'directory plans', were not really physical plans, but a series of proposals for regional development. In 1966 central government acknowledged the need for greater intervention in town planning. Local authorities were not active, so the central government initiated the preparation of subregional plans for the forty-eight subregions of Belgium. The task of preparing these plans was undertaken by university departments or private practitioners under contract to the government. The objective was to prepare plans that were sufficiently detailed to compensate for the lack of local plans. Gradually draft plans were submitted to the municipal councils, then submitted to public inquiry, and are now in the process of receiving royal ratification.

In view of the detail contained in these plans local authorities still limit their town-planning activities to the preparation of special plans and authorising land parcelling or building permits. A few local authorities have commenced preparing structure plans, but the experience of this is too limited to draw any conclusions yet.

The overall organisation of planning in Belgium has a highly centralised structure both in policy-making and procedure. Devolution to three regional authorities, which has been under discussion since 1970 (and became *de jure* in 1980), has not changed this principle. Instead of centralisation around one national authority, there is centralised control by each of the three regional authorities. Provincial authorities have no role in town planning, and local authorities are subject to strong central power.

THE STYLE AND PROBLEMS OF PLANNING

Advocates of planning in Belgium face several problems. Most citizens are not opposed in principle to planning as a rational activity in the public interest, but they are opposed to one of the most crucial consequences of its adoption, namely, the prohibition of freedom to build anywhere, and of the right to sell every parcel of land as a building-plot.

A glance at a map reveals street-villages and widespread ribbon development of homes, shops, cafés, farms, or factories alongside every road. Much of the country is flat, removing any topographical impediment to such development. Small farms with the farmer living on or near his land, has been the typical agricultural form. Even after industrialisation, the rural population remained high. Laws in 1866 and 1896 required the railway companies to provide cheap weekly tickets for industrial workers, to promote commuting rather than a mass migration to the cities. Large settlements grew around stations on the main railway routes, and with distances being relatively small, the population in urban occupations remained widely distributed.

Successive occupation by other nations prior to national independence in 1830, the liberal voting system imposed then, and then very recent adoption of a general right to vote, have left many Belgians with the attitude that they cannot exert much influence on the policy of central or local government, and that the policy-makers are outside the realm of the common citizen. Thus, there is also great scepticism about the possibility of participating in the planning of local communities.

Public housing policy is executed by a number of social housing corporations, which frequently have local politicians as their managers. Since there were until 1977 around 2,700 municipal authorities, this link meant that the construction of social housing was spread among many different villages and towns of all sizes, for political and diplomatic reasons, and without consideration of accessibility to employment or service centres, or to existing social infrastructure.

Land speculation has always been widespread, and almost a national sport. Local politicians as well as landowners are often active in land speculation. The extension of planning, and publication of plans, inevitably curtails this activity.

As in many countries, planning suffers from the lack of co-ordination between different departments of central government, and also by the extent of central controls over provincial and local authorities. A typical example of this lack of co-ordination, which causes difficulties for anyone attempting comprehensive local planning, is the fact that urban renewal is almost exclusively dominated by the Culture Department, and that the construction of housing is not seen as a part of planning policy, but as a separate aspect of social policy. Town planning officials are rarely involved.

In spite of these various problems, the concept of planning has been accepted in Belgium, and there are several examples of good practice. Many take the form of model schemes implemented prior to, or independently of, the law of 1962. For example, the concept of a garden city was realised in the development of garden neighbourhoods in a number of towns. They tended to be small, separated from the town, forming a very distinct settlement. In some cases the open green areas in the middle of the developments subsequently was used for a school or other public building, since the land was already in public ownership and the politically difficult procedure of expropriation unnecessary. In this manner the original garden neighbourhood concept was often lost.

Architects and planners often proposed innovations based on concepts of planning being developed in other countries, but these ideas were often not pursued since their significance was not always recognised by the political decision-makers. An early example is the international competition for a plan for the left bank of the River Scheldt in Antwerp in the early 1930s. Worthwhile schemes for a new town were submitted, but during the following thirty years the Antwerp authorities allowed the development of a haphazard and structureless residential area. Only in the last fifteen years has any kind of overall policy been pursued, to develop some kind of structure for the area, but it remains a somewhat unplanned neighbourhood.

In the postwar period increasing incomes, prosperity and car ownership led to widespread suburbanisation. In most villages land could be freely parcelled out and streets built in all directions. Most suburbs still have haphazard street patterns that cause traffic planners great difficulties. Separate routes for pedestrians or cyclists are very rare. Most residential areas, both recently and throughout the last forty years, have been built without any segregation of traffic, a planned hierarchy of roads, or network of routes, or any planned location of community facilities and services. The provision of efficient or convenient public transport in these circumstances is almost impossible.

As mentioned above, land-use planning proceeded piecemeal, on the basis of parcelling and building permits.

In the period of rapid economic growth after 1950 central, regional and local authorities were frequently confronted with the need to allocate industrial sites, and it was often very difficult to find one that was not surrounded or restricted by relatively new housing or a school. At this time, a number of economists advocated economic planning in Belgium, and within the universities the concepts of regional development developed by Isard and of growth poles developed by Perroux were promoted. Christaller's central-place theory was also revived in the Belgian context. A number of pioneering 'urbanist' architects and senior government officials were influenced by these ideas, and became convinced that Belgium must not lag behind other industrialised countries in adopting some machinery for planning.

Local authorities were not receptive to these ideas, so for tactical reasons regional planning was advocated in the first place, supported by central government. Starting in 1958 project groups were established to prepare regional studies for the twenty-five regions. These groups included people from different disciplines and were commissioned from private consultancies, regional organisations and the universities.

Since this period the public response to planning has progressed in two ways. At a professional level people in related professions acquired an appreciation of planning, and educational provision for planning qualifications increased. Among members of the general public, concern about the need for a better living environment increased. This was in part a consequence of the activities of the regional-plan project groups, who collected a great deal of information and held discussion sessions with all kinds of organisations, including local authorities and groups of individual citizens.

The regional studies came to fruition in the years 1964-6 but it was not legally possible for them to take the form of a statutory regional plan. They were influential studies, however, with important consequences. The end-product of the studies was a series of regional 'directory plans' based on new data, providing guidelines for regional development. Many people became aware that land-use policies, and decisions on the spatial structure of communities, are necessarily the result of far-sighted discussion and democratic political decisions. In reaction to this a number of local politicians realised that democratic planning could mean the end of easy land speculation, and a number of executive officials and senior advisers of the responsible minister felt that planning would erode their centralised power. Consequently, as well-intentioned officials started drawing up subregional land-use plans from 1965 onwards, it took a long time before any achieved royal approval. In the period 1978-80 this was achieved by the plans in Flanders, but very few of those in Wallonia are approved.

In spite of the progress to date, these subregional plans are not by themselves a satisfactory basis for a good planning policy. Although initially based on the multi-disciplinary regional studies, the value of the plans was reduced as a result of the many compromises between pressure groups, officials and politicians at all levels of government. The plans have no timescale, and no revision is provided for. The definitive drawing of the approved version of the plans was done not by the planners who prepared them, but by central government officials in Brussels, who in general lacked the necessary local knowledge. This is very unsatisfactory, as the plans are drawn with reference to the requirements of the Land Registry rather than in response to actual local conditions. These plans designate a land use for every parcel of land. This level of detail conflicts with the strategic logic of subregional planning and deprives the local authorities of scope for further initiative. Those local authorities which are planning and implementing their own policies, for instance, for urban renewal, have to devise their own techniques and operate independently of official government policy. The government minister responsible for planning is advised by members of his staff, usually junior and inexperienced politicians from his party without training in planning, and this is often reflected in official policy. There is also the tendency observable in many countries for different government departments to fail to co-ordinate their policies and programmes in a comprehensive way.

Since the revised legislation of December 1972 sub-regional plans have simply indicated possible land uses for each plot of land, broadly allocated between different categories of housing, industry, service and commercial uses, agriculture, recreation and other uses.

As an example of the subregional and structure-plan process, the preparation of plans for the Gent subregion is described. In 1964 a project group from the State University of Gent completed a regional planning study of the province of East Flanders which identified a number of options for the further development of the province. These options had to be made more specific in a number of plans for the different subregions of East Flanders. In the case of Gent this work started in 1966.

An interdisciplinary planning team, including two architects, an engineer, an architect-planner, a regional economist, a landscape architect and an economic planner, was set up. A number of other specialist advisers were called upon. The group formed a cohesive planning team.

The Gent subregion is an important part of the Rhine–Meuse–Scheldt delta, comprising the Gent area itself, the zone around the canal link to the Scheldt and a number of smaller centres. The plan had to achieve a compromise between the need to protect the remaining areas of open countryside, and the need to take advantage of the potential for development. The approach was to start at the strategic scale, and move towards more detailed work. Detailed studies of land use, transport, problems of inadequate housing or derelict industries, open space, recreation and landscape were undertaken along with a study of the regional economy. Policies for the rural areas, industrial and port areas, roads, public transport and housing were prepared.

The process of plan formulation and approval was quite long drawn out. By 1967 a broad plan on 1:50,000 scale was prepared, which was the subject of extensive consultations with local authorities, public services and other organisations. Large numbers of people representing farmers, industrialists, traders, youth groups, churches and cultural or other associations participated. After this stage, in the period 1969–72 the plan was revised and presented at 1:10,000 and 1:25,000 scales. After this stage, responsibility was taken over from the project group by the appropriate government department, who held further consultations with local authorities in the period 1972–4. In 1974 ministerial approval was given to proceed to formal public inquiry. The plan was placed on deposit, and an inquiry was held in 1975. Following this, all comments and representations were considered by the regional advisory committee, who held hearings into several thousand objections during 1976. Finally, the committee reported their findings to the minister, and the plan received royal approval in 1978.

In recent years a number of planners in both public and private sectors, critical of the disadvantages of the existing subregional planning process, have developed the concept of structure planning. This involves the democratic examination of options at a local level, modifying if necessary the provisions of the subregional plan, and setting out a programme and timescale for implementation within a local policy plan. Although this type of plan is not statutory, a number of local authorities have adopted this approach and have had to overcome administrative obstacles from provincial and national officials in so doing. These experiments are very interesting and are being supported by a number of local politicians who want local planning to be more flexible. The attitude of the three regional governments, and the overall success of this approach, cannot yet be ascertained.

PLANNING AGENCIES

Planning departments exist within both central and local government, but at the provincial level there are normally only small legal departments handling appeals from individual citizens on building or parcelling permits.

At the central government level, prior to the regionalisation of

Belgium, planning was the responsibility of a part of the Ministry of Public Works known as the Office of Urbanism and Regional Planning. It comprised a number of central departments, plus offices in each provincial capital. These provincial offices were the normal point of contact for local authorities when preparing their planning schemes. Proposals for general or special plans were submitted to the central office for a decision by the minister. Now the central office is split into three parts, one for each region, and each part is under the respective regional secretary of state for regional development and planning. As mentioned earlier, the ultimate level of authority is the Cabinet of the secretary of state, where even a building permit may be allowed although the proposal is in conflict with the approved plan.

On a local level most local authorities have a contract with an approved architect or 'urbanist' (see below), who acts as consultant to the authority, draws up development plans and advises on requests for building or parcelling permits. The administrative work is done by an official of the municipality. A number of larger towns have their own planning department, employing their own urbanists as officials in charge of permits and the preparation of special plans. With rare exceptions, such as the authorities adopting the structure-plan concept, these planning departments do not engage in overall planning of the town as a whole, but only in specific schemes.

'Urbanists' in Belgium are generally architects who include within the scope of their practice some urban design or planning work. Training as an urbanist is given as a supplementary course in a number of architecture schools, but it is essentially oriented towards physical urban design.

In recent years a few purely planning practices have been established, and since the 1960s postgraduate courses in town and country planning, on an interdisciplinary basis, have been offered at a number of universities. These courses normally comprise one year of study, plus the preparation of a dissertation. They have a rather different orientation from the architects' urbanist courses, being more directed towards the development of planning as a way of thinking and as a procedure for considering the spatial impact of decisions at every level of authority.

These differences of approach and end-product are a major source of discord between urbanists and planners in Belgium. It is noteworthy that much of the stimulation of ideas and pressure to convince people of the necessity of new approaches to planning in a democratic context is coming from the planners. The Flemish Federation of Planners was set up on the initiative of postgraduates from Gent University, and is trying to co-ordinate and, it is hoped, eventually unite the different associations of urbanists and planners, so that they can work together for better planning.

LEGAL AND CONSTITUTIONAL BASIS

The first legislation for town and country planning was the Act of 1915 by the Belgian government-in-exile in Le Havre during the First World War. This law was concerned with the postwar reconstruction of towns and villages, and provided for central government to take over responsibility for reconstruction plans drawn up by a local authority. As local politicians preferred to remain independent, no schemes were drawn up under this law.

During the period 1920–40 it was proposed to legislate for a definitive Town and Country Planning Act. However, after lengthy discussion, the proposals were amended, then abandoned, resurrected and discussed again, but never voted upon. In the summer of 1940, after the invasion of Belgium, the government fled to France and then to London. National administration in Belgium was carried out by a committee of secretaries-general, who on 12 September 1940 issued a decree on town and country planning, which was a synthesis of the proposals discussed for so long by Parliament.

Thus, a rudimentary town-planning system was started, but it remained primarily an administrative procedure concerned with building permits. After the war, the decree was ratified by the regent on 2 December 1946, and a Town Planning Office was set up within the Department of Public Works, which also had responsibility for roads, waterways, and so on. The definitive Town and Country Planning Act was finally adopted in 1962, and subsequently revised by two enactments in 1970. These represent the latest general legislation for planning.

Since 1970, however, discussions on devolution and regional reforms have been underway, with a view to administering the country in three regions: Flanders, the Brussels agglomeration and Wallonia. This reform will involve restructuring central government, and a number of sectors of government will become the responsibility of the three regional governments. National legislation in these sectors may be replaced by regional decrees. In 1980 planning became a regionalised matter, and consequently different legal systems could emerge in the three regions.

In the meantime the law of 1962 is still the basis of town and country planning, and the following principles that it embodies still apply. The whole planning system - including plan-making, the issue of permits and enforcement - is the ultimate responsibility of the central authority. This now means the appropriate regional administration. The local authorities are responsible for making local plans, and the central authorities make regional or subregional plans. Provincial authorities have only an advisory role. All local authorities should make a general plan, but this is not compulsory for authorities with under 1,000 inhabitants. Planning is a concern of the local community, and extensive local public participation must take place. Local plans can be a basis for compulsory purchase or land consolidation. Local building regulations must conform to national regulations. Local authorities issue building or parcelling permits, and appeals against refusal may be made to both the provincial authorities and the minister.

The activities of the central and local planning authorities are carried out with advice from three different types of specialist committees. The first of these is the National Committee on Town and Country Planning. Its membership consists of a president and twenty-seven members, nine proposed by the Central Economic Council, nine by the provincial deputies and nine by the minister. Its task is to propose general guidelines for the preparation of plans, and to advise the minister on new ideas and principles of planning, the designation of areas for which plans should be drawn up and on any other matter raised by the minister. There is also a Committee of Experts, consisting of three urbanists and representatives of the Minister of Public Works and the Minister of Finance. This committee advises on proposals for compulsory purchase associated with the local plans. Within every province there is a Regional Advisory Committee, which advises on the local plans proposed by the local authorities and considers objections to the provisions of subregional plans, as was noted above in the case of the Gent plan. It is probable that in the next few years local authorities in Flanders with over 5,000 inhabitants will be required to submit to the Flemish regional authority proposals for a local advisory committee.

CONCLUSIONS

Town and country planning in Belgium is still not wholly accepted. There are still politicians with a vested interest in maintaining the freedom to make deals in the absence of development plans, and it is still necessary to argue that planning has a place in a democracy and that a good spatial structure of development is in the public interest. Of course, many politicians and government officials do support the concept of planning, but this poses another problem. Although Belgium has now enjoyed 150 years of independence, there is still a feeling among many citizens that they lack any power to influence public policies, in spite of the opportunities to do so provided by planning legislation and acknowledged to be important by planners. The feeling expressed in the Catholic proverb *Roma locuta, causa finita* ('If the prince speaks, the cause is finished') frequently still prevails.

Nevertheless, it is satisfying to note that in the last fifteen years more people are becoming interested in planning, and that a number of local authorities are innovating with the new structure-planning concept. It is to be hoped that these innovations and experiments in planning will be carried out by suitably qualified people and lead to satisfactory results, so that the concept of planning will be promoted.

Chapter 7
Luxembourg

N. von Kunitzki

REGIONAL PLANNING IN A MICROCOSM

Analysing the legal and administrative structures underlying planning in Luxembourg, one has to keep in mind two basic facts: first, that the country's dimensions are exactly 1,000 square miles (2,586 square kilometres) and that its population amounts to about 350,000, the population of a suburb of London, Paris, or Cologne. Elsewhere such suburbs are normally administered from above, their planning problems being solved by a central administration. However, Luxembourg's population (or rather the three-quarters of the population that are Luxembourg nationals), has to take decisions in all matters, with absolute sovereignty.

This, of course, presents the country with a double difficulty. First, there is undoubtedly a problem of competence, although this may be overcome in the most critical and important instances by resorting to foreign advisers, consultants, or entrepreneurs. The second problem is more difficult to handle. In practice, all problems have to be solved by the people directly involved – vested interests, personal opinions, individual preferences and jealousies influencing or blocking solutions make any logical and objective answer difficult to attain. Emotion and political pressures reserved in bigger countries for greater political matters – affect any decision about a parish road or the placing of a waste outflow.

Again, this sovereignty in microcosm is of recent date, unlike, for instance, the Swiss cantons, which have had a long time to develop a mature sense of government in spite of their small dimensions – which are comparable to those of the Grand Duchy of Luxembourg.

Sovereignty for Luxembourg began, in practice, in 1867, when the Great Powers at the Congress of London decided that the fortress of Luxembourg, the 'Gibraltar of the North', should be demolished in order to eliminate a traditional bone of contention between Bismarck and Napoleon III, and which threatened to be the pretext for the imminent Franco-Prussian War.

Up to then, foreign powers had taken all decisions in the field of planning. These decisions were mostly, if not exclusively, guided by the strategic importance of the fortress of Luxembourg. The surrounding country was ignored; the economic, social, or cultural development of its inhabitants was of little concern to the successive Spanish, Austrian, French and Prussian masters of the citadel.

In 1867 the town of Luxembourg itself had little over 12,000 inhabitants, and no other town reached 5,000. With the exception of a few manufacturers in the suburbs of the fortress (which used to be entirely burnt down during every conflict), there was no functional diversification in the country of Luxembourg, agriculture being spread evenly over the 1,000 square miles. Needless to say, no rules at all restricted the liberty of private houseowners or entrepreneurs from establishing themselves on their own ground along the highways or on the gentle hills of the north. Only the fortress of Luxembourg itself had very complicated regulations restricting the right of establishment, due only to considerations of security and military convenience.

Industrial development, which came late to Luxembourg compared with the industrial regions of Britain or continental Europe, also developed freely, without any government or even local intervention. As industrialisation hit Luxembourg much more abruptly than elsewhere in Europe (latecomers advancing normally more rapidly than forerunners), the Grand Duchy did not experience the massive rural exodus to the new industrial centres. Lack of time forced entrepreneurs to import foreign labour more familiar with an industrial environment. During 1870-1910 Luxembourg imported about 40,000 foreigners, or more than 15 per cent of its total population, the highest percentage in Europe.

In the industrial part of the country this percentage was, of course, much higher and accounted for most of the additional population in the south of the country. Consequently, town and regional planning in the south was mostly done by the private entrepreneurs who had to procure shelter and food for the new populations. A certain number of these entrepreneurs came from abroad, mostly from Germany, since Luxembourg adhered to the German Customs Union during 1842-1918. In so far as the industrialisation of the country was accomplished by the Luxembourg bourgeoisie they largely relied on German specialists for managing the steel companies and other industries that flourished in the wake of the rise of Germany as an industrial power. It was the steel companies that planned the industrial suburbs around the blast-furnaces and steelworks, and built hospitals, schools, roads and power stations. Even today the steel company which employs over 90 per cent of the steelworkers, and more than a quarter of the industrial population, owns enormous reserves of land in order to be able to develop its long-term investment strategy, with the social developments involved.

When Luxembourg built its railway system, the secondary network was constructed by private foreign entrepreneurs who agreed to build the whole infrastructure against allowance of mining rights in the south of the country. As these mining rights were conferred proportionally to the track kilometres built, some of the railways took the most baroque form, undulating through the country with funny arabesques just in order to allow the steel companies to construct a maximum-length railway between two stations, fixed by the government. This example sufficiently illustrates the absence in the late nineteenth century of any real concept by government of planning principles, when Luxembourg made its first steps as an independent nation 'planning' for itself.

THE STYLE OF PLANNING: URBAN DEVELOPMENT IN A FOREST

When the troops of the French Republic invaded the Austrian Netherlands and took over power in Luxembourg, they named it Forest Department. There was little else in 1795, apart from the capital contained within its centuries-old fortifications. Urban problems were unknown to the population, responsibility for any local problems residing with elected aldermen; and the few roads had always been planned and maintained by foreign governments. Therefore, when Luxembourg, seventy years later, was confronted with the problem of demolishing the fortifications and planning a modern city, it had the excellent although obvious idea of consulting a French architect, M. André of Paris. André's design in seven years transformed a mediaeval fortress into a modern city, where concentric boulevards surrounded the old nucleus bordered by 50 hectares of parks, planted on the waste from demolition works.

André's task was largely facilitated by the fact that, for once, it was national and not local authorities that directed the works. The fortress was a national concern and so Vannerus, Director of Public Works, directed the plan. Furthermore, for military reasons, the state was the owner of the territory immediately surrounding the fortress, so that, for the first time in Luxembourg's history, the government was able to develop an urban plan in the years before the First World War, as far as Luxembourg City was concerned. For the rest of the country, the only rule concerning house-building was one imposing alignment along the sidewalks, a regulation first developed by the city of Luxembourg in 1873 and imitated by other towns of any importance.

At the end of the 1870s the steel industry developed in the south-west of the country, thanks to Thomas Gilchrist's improvement of the Bessemer process. The rural population of the east and the north began their migration towards the industrial south, where several agglomerations of 10,000–30,000 inhabitants developed during 1865–1935. The population of Luxembourg's purely rural districts was reduced from 55 per cent of the population to slightly over 30 per cent, while the south increased from 8 to 33 per cent of total population.

While private companies showed a certain sense of national, or at least regional, responsibility in building their factories and social institutions, general house-building followed no other rule than the one of alignment mentioned above. Thus, towns spread out enormously along public roads, so that most of these agglomerations show an underdeveloped centre and very unaesthetic, unfunctional, ribbon development through the countryside.

This unfortunate development fully illustrates the difficulty of planning in a microcosm, where too many private vested interests come into conflict with general objectives. The case of the city of Luxembourg, however, shows that a lot more could have been achieved had initiative and control not been confined to local authorities, but brought to a national level, a level low enough in view of Luxembourg's dimensions.

This lesson was slowly learned and politicians of all colours as well as public opinion became more and more convinced that urban and regional planning in the second half of the twentieth century was too important to be left to the competence (or more often incompetence) of local authorities. The Minister of the Interior elaborated a law revising the existing division into 128 local communities (communes), replacing them with thirty-nine bigger units, able to support an administrative staff, and thus to solve the many questions which the present legislation leaves to the local authorities. Unfortunately, Parliament could not muster the courage to make this compulsory (many members were also local mayors), and instead, tried to persuade the local communities concerned into voluntary mergers, a strategy that has had very little success. Only two mergers have taken place on this voluntary basis, as once again, local jealousies and personal ambitions blocked progress along the road of functional logic.

In order to be able to tackle the problems created by rapid economic and social development, as well as industrial evolution since the Second World War, and especially since 1960, the government was forced to resort to government agencies. These are established by special laws which not only grant them the funds necessary for their task, but also grant special status and powers exceeding the normal degree of intervention of government into matters of local concern. They operate under the direct supervision of the appropriate minister.

The city of Luxembourg is an outstanding example of the dramatic change undergone by urban infrastructure in Luxembourg. Up to 1955 Luxembourg City could be considered a provincial town, to which the status of a national capital gave some importance and style, but which lived out of the mainstream of political, social and cultural evolution. As social progress moved forward in Luxembourg as elsewhere in Western Europe, pressure on the general infrastructure increased accordingly, but in the case of Luxembourg City this development was dramatically increased by the new role of Luxembourg as an important administrative centre of the EEC and the most important financial centre on the continent, containing the European Investment Bank. These two additional factors have posed for the city fathers a problem comparable to the one that had to be solved by the authorities in 1867. And not surprisingly, the city council now turned to a French architect, Pierre Vago, who in 1964–7 developed new building regulations. M. Vago had no authority or independence of his own, but acted as a consultant for the city council, and few of Vago's plans for fundamental innovations have been realised.

The coexistence of very rapid economic, social and cultural development alongside the nineteenth-century administrative divisions, and building regulations applying to a purely rural environment, mean that the national government and the Council of the city of Luxembourg face the urgent need to formulate new regulations. These are required to plan for necessary industrialisation, and for Luxembourg City's new role as a European service centre, while respecting a public desire to keep the Grand Duchy as 'the green heart of Europe'. As a centre of tourism, a haven of beauty and calm, Luxembourg must grant legal security to its citizens as well as to those foreigners wishing to establish themselves there. Public opinion has become very conscious of the conflict of interest that can arise between these three sets of considerations. After almost a century of inactivity in the legal field, Luxembourg's authorities are actively preparing the legal and administrative network that will sustain such a development.

THE LEGAL PATTERN

The first concrete basis of town planning in Luxembourg is the law of 12 June 1937 concerning building and planning of cities and other major agglomerations. This law, which took ten years in passing by the Luxembourg Chamber of Representatives (1927–37), went further than just prescribing the alignment of houses, as had legislation existing up to then. It forced municipalities to establish a street plan and adopt development programmes for general services, such as water, energy distribution, drainage, and so on. The plans provided rules concerning hygiene and archaeological or aesthetic limitations for different city areas. However, the law of 1937 only applied to cities above 10,000 inhabitants or to towns of special picturesque, artistic, or historical interest. The obligation to present an official plan was also extended to private associations, companies, or persons who were about to create new agglomeration.

Whenever any existing built-up area is destroyed by fire or natural catastrophe, the law provided that it can be rebuilt only after a town-planning scheme has been authorised by the government except by special authorisation of the town council and competent minister. A delay of three years was granted to local communities falling under the law to comply with its prescriptions, but unfortunately this stay coincided with the German invasion in May 1940, and so the law remained mostly ineffective.

The destruction of about one-third of the country during the Rundstedt offensive in the winter of 1944-5 offered an ideal, though unwanted, occasion to reorganise the whole rural north of the country. But in their understandable eagerness to ease social suffering and encourage any private initiative contributing to repairing war damage the Luxembourg authorities missed the occasion to enforce the law of 1937, as far as it referred to areas destroyed by cataclysm. Not only were local communities freed from submitting an urbanisation plan before rebuilding their individual area, but the authorities also failed to reorganise the whole administrative structure. The rural areas were still regulated by the laws of the French Revolution, imposed in 1791, creating the rules of the Département des Foréts, and transforming each parish into a commune. Since then only the capital had been reorganised, in 1920, by merging four major suburbs into the territory of the commune.

All over the northern area, devastated by the Rundstedt and the Allied opponents, schools, churches and administrative centres were rebuilt where they stood, in spite of the fact that the population had long before declined greatly. Twenty years later legislators tried to repair the errors and omissions of 1945, with little success.

After the Second World War, the government started an industrial diversification programme that succeeded in attracting a certain number of new industries to the less industrialised parts of the country, but these important cases were rare enough for the government to solve the problems on an individual basis, through a pragmatic co-operation between the different departments concerned. As the steel industry gradually ceased to be as a powerful force for development a new diversification programme was aimed at encouraging entrepreneurial initiative on a more general basis. On the other hand, the role of Luxembourg as a European administrative centre created new needs and problems.

One of the special laws mentioned above, that of 7 August 1961, created a Planning Fund for the construction of a European administrative area based on Kirchberg to provide facilities for the European communities established in Luxembourg. The fund is administered by a planning board under the authority of the Minister of Public Works. A law of 16 August 1967 established a Roads Fund, also administered by the Department of Public Works. The law in addition laid down rules of expropriation in relation to the construction of roads.

Meanwhile, a law of 25 May 1964 tried to repair the omissions of 1945 by organising the redivision of agricultural holdings, under the authority of the Minister of Agriculture. The purpose of the 1964 law was to form a rational units of agricultural production, adapted to modern, capital-intensive methods of production and taking into account the diminished number of landowners. The Ministry of Agriculture divided all the rural land into production units; the landowners within such units formed syndicates, authorised to decide whether to divide their estate or not. A national council presides over the redivision of the rural estates, and this has been quite successful especially in the north of the country.

For all these reasons, a legal framework for all aspects of planning has been developed since 1960. Understandably the different laws and regulations have been adopted in somewhat dispersed order, as problems arose. Contradictions were inevitable and a need for harmonisation was gradually discerned. This led to the law of 20 March 1974, which is the backbone of the present legislation concerning town and country planning: the *Loi concernant l'aménagement général du territoire*, which for the first time in Luxembourg's history aims to establish a general planning system, including all economic, social, cultural and environmental aspects.

The law of 1974 created the new post of Minister of General planning, traditionally entrusted to the president of the government. The Minister of Planning is assisted by a High Planning Council, comprising mostly representatives of associations and organisations interested in certain aspects of country planning (the Order of Architects, Economic and Social Council, environmentalists and cultural associations), and by an interministerial committee, composed of high-level public servants representing the different ministries. The committee is in charge of co-ordination of the activities of the respective ministries concerned in country planning.

This law supersedes that of 1937, in that every local community has to submit a plan for all its territory within three years of the government fixing a general programme of country planning. This was done, by the government in council, on 11 November 1977; thus, every one of the 118 communes was required to have submitted a local plan by the end of 1980. In fact, only thirty-eight of the 118 communities had submitted such a plan to the minister by 31 December 1980. The submission period can be renewed once at the request of the local authorities, and now expires at the end of the 1983.

The 1974 law also requires the government to adopt a general programme of planning, which co-ordinates the different objectives mentioned above. Based upon this plan, the Minister of Planning can submit to the government sectoral or general territorial plans. Local authorities affected have to be informed and consulted.

The general programme of 11 November 1977 contained a list of thirty-five sectoral programmes, identifying the ministry which is to take the lead, and which others have responsibilities in the relevant sector. Themes include public works, infrastructure, transport, education, industry, tourism, environmental quality and urban conditions.

The range of responsibilities, often overlapping, makes for inefficient planning. The industrial responsibility of the Ministry of National Economy, for example, needs to be co-ordinated with that of other ministries in energy, transport, and so on. Therefore, a decision of the government in council, of 4 August 1978, adopted a geographical plan creating industrial zones in the south of the country. This geographical plan at least partially realises the industrial programme included in the general programme of November 1977.

The outline of the legal pattern on which planning in Luxembourg is based would be incomplete without the environmental aspects regulated, for the first time, by a law of 29 July 1965. This law, concerning general aspects of protection of nature and national resources, also ruled that building was not allowed outside any agglomeration of houses without the consent of the Minister of Forests and Waterways. An agglomeration is defined as five houses inside a radius of 100 metres. Stricter control came with a law of 27 July 1978, which absolutely forbade any domestic building outside an agglomeration. Even the Minister of Forests and Waterways could only allow the construction of agricultural, commercial, or industrial buildings outside the agglomerations. This law could normally be expected to compel the minister to reinforce the existing patterns of development, along roads where five or more houses already exist; although it might prove more satisfactory, for overall planning reasons, to control the development of existing concentrations, and to promote the creation of new communities.

A proposal for a new law has, therefore, been submitted by the present government, based on the law of March 1974 concerning general planning, and on the general programme of 1977. If the new law is passed, any building will only be allowed inside a zone defined by a plan established by the local community and authorised by the Minister of Forests and Waterways. The new law aims at co-ordinating the legislation concerning country planning and that protecting the national environment, while eliminating the irrational and damaging restriction on the extension of existing agglomerations.

Unfortunately, as has been mentioned above, only a small fraction of the local communities in Luxembourg have, until now, complied with the law enforcing the establishment of a local plan. Until these plans are prepared, building authorisations have to be granted by the minister on a case-to-case basis.

PERFECTION AS A HANDICAP: THE COMPETENCE MAZE

As has been shown, planning proves rather difficult in the Grand Duchy in spite of, or perhaps because of, its small dimensions. At the beginning of the chapter we referred to some of the factors that contribute to hampering effective planning in a very small country, where individual cases are not submerged amid great numbers and where every single proposal arouses political opposition. Important cases, where legislation, rational in itself, has not come to realisation, have been illustrated.

One additional difficulty which rather impedes progress in Luxembourg is the ambition of the legislators to establish all the legal competence that can be mustered. This is to be expected in a small country, where there are no universities and no scientific foundations or research institutes that can be charged with a thorough analysis of individual cases or programmes. The different ministries, while being rather well endowed with legal competence in their own field, are not assisted by any multidisciplinary research teams able to work on any subject which has many facets as urbanisation and country planning. However, the Luxembourg authorities are deeply impressed by the need to consider all aspects of country planning. The law of 1974 explicitly states that all aspects of human life should be considered and taken into account by the different plans. The multidisciplinary nature of planning – and the range of disciplines involved – is also explicitly acknowledged.

Unfortunately, the present legislation appears to be profoundly marked by this perfectionist mentality. In its comments on a proposed law concerning the elimination of waste, the Chamber of Commerce recalls that for a new industry trying to establish itself in Luxembourg, the authorisation imposed by the new law would come on top of seven other authorisations presently necessary in order to establish a new industrial activity anywhere in Luxembourg: an authorisation of establishment by the Ministry of National Economy and the Middle Classes; and authorisation for dangerous, insalubrious or incommodious establishments, delivered by the Ministry of Justice; an authorisation concerning environmental aspects by the Ministry of the

Interior; a building authorisation; an authorisation *commodo et incommodo* by the local authority; an authorisation to use the public roads by the Department of Roads and Bridges; and an authorisation of access to the national railways.

Of course, these multiple authorisations impede the establishment of any new industry or other venture: if an applicant is not advised by an expert, he risks losing time and money through delays due to the failure to seek one of them. A much greater obstacle to harmonious, rational and efficient planning of industrial and social development is created by the fact that the different authorities whose permission is necessary for any new establishment, are entirely autonomous in their decision, one being able to block all the others.

As Luxembourg has been pursuing, since 1949, a deliberate policy of attracting new industries (mostly subsidiaries of US multinationals) to the country, this handicap is largely overcome, in the case of potential newcomers to Luxembourg, if they are considered to fit the industrial diversification policy. In these cases, a working party of the Departments of Economy, Middle Classes and Finance takes charge of the applicant and obtains all authorisations necessary, acting as a representative of the applicant.

No such facility exists for the poor individual or the small investor who tries to establish himself in Luxembourg or, simply, build a house outside the normal agglomeration. This, of course, inspires a high degree of legal insecurity for the normal investor, and doubtlessly deters more than one private initiative. In other cases many entrepreneurs and individuals simply ignore the law; individuals just proceed with their investment, hoping that the complex legal procedures are such that they can complete their project before any reaction comes from the authorities. If eventually such a reaction does come, more often than not triggered off by protests of environmental associations, a *fait accompli* is presented which the legal authorities hesitate to undo, because of the severe financial consequences.

To summarise a complicated situation simply, one might say that the intricate legislation and combination of competences in Luxembourg's town and country planning regulations prove a big handicap to the law-abiding and an equally strong temptation to those who do not think the law respectable on this point.

There is a clear need to have one administrative department responsible for all aspects of a project presented by the applicant for a building licence, the authorisation to establish a new industry or other venture. Such a department should be able to prevent any initiative starting illegally, in good faith, some important authorisation having not been obtained, due to the applicant's lack of information; and to ensure that the different authorisations are obtained in a reasonable period, so that the project is not delayed by some administrative department. It should establish clear procedural channels along which a proposal runs. Presently it is possible that several administrations pretend that a project is retained in another place, while the applicant is not able to ascertain where his application is delayed. It should act as an arbiter if any administration opposes a project on grounds of its competence, while the same project seems worthwhile to other departments. Normally only the president of the government, who traditionally administers also the Ministry of General Planning, is able to play such a role.

Unfortunately, at the moment, the Ministry of General Planning is practically reduced to a small secretariat whose task is one of co-ordination rather than of analysis and decision. For reasons which have been discussed, it is necessary to use competences existing in the different specialised ministries rather than to establish a parallel hierarchy inside the Ministry of Planning. The latter is not an option, for financial reasons as well as because of a lack of suitable people, but the role of the different ministers with planning responsibilities should be one of consultation. The final decision can only be taken in one ministry, on the basis of the programme of general planning which has been established by all the ministries in co-operation.

The Luxembourg system of project authorisation ignores the fact that, almost inevitably, any individual project contains advantages and disadvantages in the different fields covered. And according to the approach chosen by the analyst, which might be financial, economic, social, cultural, environmental, political, military, or whatever, advantages and disadvantages prevail. Only an analysis covering all these issues can make a rational trade-off between, for example, economical advantages and environmental drawbacks.

In the present system of planning an inevitable stalemate is quite clearly built in. If it has not been more evident in the past, it is only because Luxembourg's small dimensions imply, besides the disadvantages mentioned above, a high degree of pragmatism, unofficial compromise and mutual understanding. Nevertheless, the existing legislation should be complemented by a law enabling the Minister of General Planning to act not as a mediator between decision-makers, but as a decision-maker himself, assisted by a number of specialised consultants.

PLANNING AS A KEY TO A NEW ERA

Effectively, the necessity of government ruling private initiative in the social and economic field has been recognised only after the Second World War. Up to 1960 state or local regulations only concerned marginal aspects of building or economic activities. This may seem surprising since Luxembourg has been from the beginning of this century one of the most industrialised, best organised and most prosperous countries in Europe.

The explanation is historical. Up to 1860 Luxembourg was an underdeveloped country, living on potatoes and on the expenses of a badly paid Prussian garrison in the fortress of Luxembourg. When industrialisation came to the country, it was organised by big firms, mostly foreign, that effectively undertook their own urban and regional planning. If the development of Luxembourg's economic and social infrastructure might not appear ideal by today's planning standards, it must be admitted that it avoided most of the social and economic errors experienced by countries developed earlier, in a wave of unorganised, individual entrepreneurial initiatives.

While the south-west of the country was thus industrialised by a few big foreign companies, the rest of the country, already sparsely populated, was slowly emptied of its population, no town having over 5,000 inhabitants. Luxembourg has a population density of 140 inhabitants per square kilometre, compared with 323 for Belgium, 344 for the Netherlands, 247 for Germany and 170 overall for the EEC. Thus, the Grand Duchy of Luxembourg was spared most of the urbanisation problems experienced by other countries.

An elaborate legislative framework covering different aspects of planning has been built up since 1960. Many contradictions emerged, which the law of 1974 sought to resolve. Competent people are difficult to find in Luxembourg outside the specialised administrations, and each government department has a high sense of its priorities and of its exclusive right to rule its field of activity. Also local communities maintain their spirit of autonomy; due to Luxembourg's dimensions, these communities are often so small that this spirit clashes with an ability to understand and to act.

Legislators, anxious to respect democracy while seeking a high degree of perfection in organising the life of the community, have created a complicated administrative system. Most of the planning is still left to the local communities who have often proved unable to produce the fundamental plan for their territory provided for in the general planning programme. As for planning on a national level, there exists a Ministry of General Planning – whose task is one of co-ordination rather than of decision. Fortunately, the high degree of coherence in a small nation makes things easier in most cases. But a higher degree of integration has to be achieved for general planning, if the very ambitious national programme, such as that established in November 1977, is to be put into practice.

As environmental demands surge at a moment when the economic future of Luxembourg appears rather bleak due to the steel crisis, the need for a straightforward policy of development based upon a balanced planning programme, respecting social, cultural and environmental needs, becomes apparent to everyone. An official programme embracing all these aspects has existed since 1978. The necessary administrative structures are there, but there are too many centres of political decision-making. It would be desirable for the Ministry of General Planning, which is the only one capable of taking balanced planning decisions, to become the overall authority for town and country planning.

Chapter 8

The United Kingdom

R. H. Williams

In order to convey an understanding of the scope and style of British planning practice it is realistic to consider the range of activities undertaken by local planning authorities, developing or changing the physical character of their area in order to improve housing, employment prospects, social and leisure facilities, transport, and so on, for their citizens. It would be quite wrong, at least as far as Britain is concerned, to equate the scope of planning practice with that of the Town and Country Planning Acts. The activity of planning, and of agencies responsible for planning, is much more wide-ranging than this. Moreover, there have been great increases in the scope of planning during the postwar period.

These have come about in distinctive phases which have occurred in response to particular circumstances of the time. The net effect has been one of steady growth of the planning system, although some recent developments under the present government could be interpreted as the beginning of a reverse trend. Recent legislation has made certain changes in the planning system, but how significant these changes will prove to be cannot yet be judged.

In spite of the many later developments, the basic features of the system date from the period of reconstruction in the late 1940s. The 1947 Town and Country Planning Act brought all land under a system of development plans and the control of development. More specific measures, providing for the designation of new towns and national parks, were also enacted during this period, and together they laid the foundations of Britain's postwar system of comprehensive planning. Very high expectations were held for the new system by people who had experience of planning in wartime conditions and who were anticipating a major role for planning in the postwar period.

The present-day significance of the innovations of each postwar decade is considered in turn. During this period there have been a number of occasions on which a new theme or approach to planning has become a major priority of government and local authority policy. Sometimes the impetus has come from a significant body of research findings, and sometimes from political initiatives or effective lobbying by major pressure groups. In general, the pattern followed by these innovations is one of initial study or experiment either by independent researchers or by a few local authorities, then the period of popularisation of new ideas, and their enthusiastic incorporation into the planning process, perhaps accompanied by legislation, followed by the adoption of the innovation as a routine element of planning.

PUBLIC AND PRIVATE SECTORS

The salient feature of British planning is the fact that it is, and has been for some years, primarily a public sector activity. Consequently, the focus of attention must be concentrated on the activities of local planning authorities and other official organisation with planning responsibilities including the offices of central government, the New Town Development Corporations and National Park Authorities. These authorities employ between them a large number of professional planners, and the majority of professional planners in the UK are employed in this way. These authorities generally expect that all the necessary technical expertise is to be found within the ranks of the salaried planning officials. The majority of these officials hold qualifications recognised by a well-established professional institute, the Royal Town Planning Institute (RTPI). There is a private sector of the planning profession engaged in consultancy work, but this sector undertakes a relatively small proportion of the planning work done on projects within the UK. Local planning authorities rarely employ consultants, who would be more likely to be advising private developers within the UK, although they frequently undertake work for government agencies of other countries. Much of British planning practice takes the form of the management of the urban environment by a body of people trained to operate in a bureaucratic environment rather than as designers of master-plans for individual clients.

STRUCTURE OF GOVERNMENT

It is appropriate, therefore, to outline the main administrative and political features of local and central government. Britain has a unitary constitution, and the power to legislate is held only by the national Parliament. Power to implement or execute policy is held both by central government and local government, and in the case of many aspects of domestic policy, including town and country planning, the local authorities are the agencies with the principal powers and duties to provide the necessary service.

The whole of mainland England, Wales and Scotland is administered at a local level by two sets of local authorities, with a structure that was substantially reorganised in 1974 and 1975. The upper level is that of the county councils in England and Wales, and regional councils in Scotland. Within England counties can be classified into three groups. London is within the Greater London Council, whose powers differ somewhat from those of other counties. Secondly, the other six major conurbations of Birmingham, Liverpool, Manchester, Sheffield, Leeds and Newcastle are each within metropolitan counties. The present government has announced its intention to abolish the Greater London Council and metropolitan county councils, but no legislation has yet been introduced. The remainder of England, together with Wales, is divided into forty-seven non-metropolitan counties: thirty-nine in England and eight in Wales. Scotland is divided into nine regional councils covering the whole of the mainland and the inshore islands. Three outer groups of islands, Orkney, Shetland and the Western Isles, are governed by a separate type of authority known as Island Authorities.

The distinction between England and Wales is of limited significance as far as planning is concerned, although it is more so in other respects. Scotland still has a separate code of law dating back to its days as a separate kingdom before 1707, and there are some significant differences of practice in planning. Northern Ireland is still governed directly by the UK government, and its planning service is run by a government department and not by local authorities as elsewhere in the UK.

Each of the English and Welsh counties, and Scottish regions, but not the Scottish Island Authorities, is divided into district councils. The powers and duties, and average sizes of districts, vary depending on whether they are within metropolitan or non-metropolitan counties, or are London boroughs within the Greater London Council. In general, the size, resources and responsibilities of London boroughs and metropolitan districts are greater than those of the other districts. There are thirty-two London boroughs, thirty-six metropolitan districts, 333 non-metropolitan districts in England and Wales and fifty-three districts in Scotland.

Local authorities are controlled by elected councillors, who do not receive a salary, and are normally not full-time politicians. There are no elected politicians in the role of full-time executive directors of a department. Executive responsibility rests with the permanent officials, including planners, and the distinction between the permanent officials and the elected politicians is maintained very clearly. The financial resources of local authorities are provided from two major sources, a block grant from central government and a local property tax, or rates, levied by the authority within its territory. This partial dependence on central government for its finance, together with the lack of power to legislate directly, limits the autonomy of local government, but there are considerable variations in local authority practice both for political reasons and as a result of different problems arising in different localities.

Political responsibility within central government for planning rests with four secretaries of state, for the Environment, for Wales, for Scotland and for Northern Ireland. The Secretary of State for the Environment, as the English planning minister, and the minister with the most specialised department, that is, the Department of the Environment, tends to be the leading influence at this level on planning matters. The whole country is divided into standard regions, one corresponding to each of Scotland, Wales and Northern Ireland and eight within England. These regions are part of the administrative organisation of central government, and do not correspond to any form of elected government, nor should they be confused with Scottish regional authorities, which are the upper level of elected local government in Scotland. The regions of England each have a regional office of the Department of the Environment. During the 1970s each region had an Economic Planning Council, but these no longer exist.

DEVELOPMENT PLANS

Although there have been major changes since, the basic principles of planning were established by the Town and Country Planning Act 1947. The first is that the whole land surface of the country is covered by a system of development plans, indicating the policy adopted by the local authority for the future development of their area. Secondly, planning permission must be obtained from the local planning authority before any development can take place.

The present types of development plan are based on reforms introduced by legislation in 1968. These are the structure plans, which are the responsibility of the county councils and Scottish regions, and the local plans, which may be prepared by counties or districts, but which are intended to be primarily the responsibility of the districts. Structure plans have to receive the formal approval of the appropriate secretary of state before they are adopted. This approval must be preceded by extensive publicity and public participation, and by an opportunity for people to register objections. A hearing, known as an Examination-in-Public, takes place before a panel of inspectors, who then report to the secretary of state recommending whether approval of the plan should be given, and what modifications might be required. The structure plan takes the form of a written statement of policy, supported by its justification and reports of surveys and data analysis that have been carried out. The policies are also expressed in graphic form in a key diagram, which does not have a conventional map as a base, and which does not, therefore, convey precise information regarding the location of proposals. The absence of a map is deliberate, since the purpose of a structure plan is to express broad policies, and avoid questions of detail better left to local plans.

The function of a structure plan is to set out strategic land-use policies for the future economic and physical development of the county, policies for the management of traffic and the improvement of the physical environment. The plan must concern itself with current policies for economic planning and development of the region as a whole, and with the resources likely to be available for implementation. The survey is a major element of the process, being concerned with economic and social forces as well as physical land-use matters.

It is intended that structure plans should be monitored, and revised when surveys suggest that this is necessary because of changing cirumstances, and some authorities are doing this. In other authorities, however, the first structure plan has not long been approved. Preparation of structure plans has proceeded much more slowly than was originally hoped.

Local plans, in contrast, can be adopted directly by a local authority, and do not require the approval of the secretary of state. Their preparation is not a mandatory duty in England and Wales, although it is in Scotland. There are three types of local plan, known as district plans, action area plans and subject plans. They are designed to provide detailed guidance on the policies of the local authority for future land use and the control of development, and detailed interpretation of the structure plan policies. For this reason, they are intended to be adopted after the adoption of a structure plan, although this is no longer a legal requirement. The policies contained in a local plan are expressed on a base map, together with a written statement of justification of the policies. The policies are, therefore, expressed in site-specific terms, unlike those of the structure plan.

Of the three types of local plan, the district plan is the most common. This is the general-purpose plan, and roughly half of all local plans being prepared are of this type. A district plan is likely to be for the whole of a small town or small district council area, or for a sector of a large town or city. Action area plans are prepared for small areas in which substantial change and development is proposed within the following ten years. Their location may be indicated, but not precisely defined, on the structure plan. The third type of local plan is the subject plan. This is prepared to establish the policy for some particular situation or theme which is capable of being handled independently of other planning issues.

The legislation which is the basis of the present development-plan system is the Town and Country Planning Act 1971, although this has been amended since, especially by the Local Government Act 1972 and the Local Government, Planning and Land Act 1980. In 1971 it was assumed that structure plans and local plans would be prepared by the same local planning authority. The two levels of development plan were expected to become an integrated package expressing consistent policies, with the local plan being adopted soon after the structure plan. However, this assumption very soon proved to be invalid. After many years of debate, local government structure and boundaries were reorganised under the Local Government Act 1972. The story of local government reorganisation is well told elsewhere,[1] but it is necessary to emphasise that the whole planning system was fundamentally affected by the reorganisation of local government which was implemented in April 1974 in England and Wales, and May 1975 in Scotland. The planning service was split between the two levels of local government throughout the country, with the exception of the island authorities and the three most sparsely populated regions in Scotland. The two types of local authority exercise their powers independently of each other, and of course may be governed by different political parties.[2] Consequently, there can easily be a divergence of policies and priorities, which may be reflected in the development plans. Therefore, the original assumption of consistency between the levels of development plan is no longer necessarily valid, and a procedure was established in England whereby a county may certify that a local plan is in general conformity with the policies of the structure plan.

In practice, the development plan system has barely become fully operational. The disruption caused by reorganisation slowed up the process of plan preparation in many authorities, and local plans have only recently come forward in any numbers. Many authorities are not adopting the formal procedures, but prefer to rely on informal plans, prepared in the normal way, but not adopted in accordance with the procedures laid down in the Town and Country Planning Acts.

CONTROL OF DEVELOPMENT

The system of development control has followed the same principles, though with many changes in detail, since 1947. The basic principle is that development is given a very wide-ranging definition, and all operations that constitute development require planning permission, unless they are explicitly exempted. The development plan adopted by a local authority indicates the policy of that authority, and therefore indicates the types of development proposal that may be expected to receive planning permission in different locations. However, the development plan does not itself create any automatic right to develop, nor does it offer any automatic guarantee that planning permission for land uses indicated would be forthcoming. The local authority, when considering an application for planning permission, must consider the development plan and any other relevant matters, but may choose to depart from the policies of the development plan in deciding the application. The decision may be to grant permission, with or without conditions, or to refuse permission. Powers to make these decisions are held by the district councils for the majority of planning applications, and by counties for major proposals which might have a significant impact on the proposals of the structure plan. The central government intervenes only in a minority of cases, either when it is intended to approve something which is a major departure from an approved development plan, or if an applicant is appealing against the decision of the local authority, or against its failure to make a decision within the specified time. When an appeal is made, an inspector is appointed by the Secretary of State for the Environment, who may hold a public inquiry, and a decision is made based on the findings, by the inspector, or in more major cases, by the secretary of state.

It is worth quoting in full the definition of development used in the Acts since 1947: 'the carrying out of building, engineering, mining or other operations in, on, over or under land, or the making of any material change in the use of any buildings or other land.'[3] This is a very wide-ranging definition, particularly with the inclusion of changes of use. It excludes demolition, as such, however, although provisions exist to control demolition of buildings of historic or conservation interest. A number of exclusions are made from the scope of development control. In general, development by government departments does not legally require planning permission, although it does undergo a similar process of consideration. Development for agricultural purposes, on agricultural land, and certain minor proposals are also largely exempt.

Consideration of a planning application is an important power of a local authority, and the discretion to decide applications on their individual merits is an important element of flexibility. However, the local authority does not decide in isolation. Widespread consultations take place, with other public authorities, or with local organisations or individual citizens affected by a proposal, and the local authority must have regard to the public acceptability of its decisions. This discretion is an important characteristic of the British system. It has advantages, but there is sometimes criticism of the uncertain relationship between the development plans and development control.

NEW TOWNS

British new towns have attracted widespread publicity. Based on the New Towns Act 1946 they have been developed by special agencies, known as Development Corporations. They are independent of the local authorities in the area, and are given the duty of planning, designing and implementing a new town. New towns were frequently designated where they could accommodate overspill population, displaced as housing densities were reduced in London and other major cities. Others such as Washington, near Newcastle upon Tyne, were designated in order to stimulate regional growth.

Early new towns had modest population targets of around 50,000-80,000 population. Later examples, in central Lancashire, Milton Keynes, or Northampton, had targets in the range 250,000-500,000 population.

Present-day policy does not give new towns the significant role they once had. The new town programme is being run down in order to concentrate resources on renewal of the inner areas of the large industrial cities. In two important respects, however, the approach to inner city revival is based on the successful experience of the new towns. One is the concept of the development corporation itself, and the other is their expertise in industrial promotion. The development corporation consists of a board appointed by the secretary of state, which in turn employs the necessary planners, architects and other staff. Unlike a local authority, this board is not subject to local political control and is geared to specific objectives of new town development. These factors enable development corporations to operate in a more direct manner than local authorities, and they are sometimes held up as models of efficiency. The Local Government Planning and Land Act 1980 provides for similar agencies, known as Urban Development Corporations, to be set up to direct the renewal of particularly run-down inner city areas. Two have been established in England, for dockland areas of London and Liverpool. A somewhat similar device is already operating in Scotland, where a central government agency, the Scottish Development Agency, is intervening directly to implement urban renewal in the city of Glasgow, having set up the Glasgow Eastern Area Renewal Project for this purpose in 1977.

The new towns have relied on attracting industry from elsewhere in order to provide an employment base for themselves. In general, they have been very successful in doing this, although some new towns such as Skelmersdale, near Liverpool, have since suffered major unemployment due to factory closures. It is widely felt, however, that this success has been at the expense of prosperity in the inner city areas, especially in London.

COUNTRYSIDE PLANNING

In 1949 legislation was passed setting up the national parks of England and Wales, to be managed by the appropriate county planning authorities, advised by a National Parks Commission. The ten parks set up have flourished, and have witnessed not only effective control of intrusive development, but also many positive planning measures to cater for increasing visitor pressure, while preserving the amenity of the area and maintaining the local economy and community. National parks have twin objectives: to preserve amenities and to provide for the enjoyment of the public. These objectives are often in mutual conflict but the national park authorities and, since its establishment in 1968, the Countryside Commission, have succeeded in overcoming this problem to a large degree. Other forms of designation more explicitly designed to conserve include Areas of Outstanding Natural Beauty and Areas of Special Scientific Interest.

The Countryside Commission replaced the National Parks Commission in 1968, with a wider brief. In addition to the advisory role in relation to national parks, the Countryside Commission has responsibilities for facilities in the countryside, to conserve natural beauty in and secure public access to the countryside. It promotes research, advises local authorities and has the power to initiate or support experimental projects.

In association with this enlarged role for the Commission, the Countryside Act of 1968 gives local authorities powers to designate country parks and picnic areas. Country parks are intended to offer planned provision for countryside recreation in suitable locations close to the main centres of population. In doing so, it is hoped that they would relieve pressure on remoter, more sensitive locations, and reduce the danger of damage to the rural environment from haphazard picnicking or other activities by town-dwellers in undesignated places in the country.

One lasting innovation from an earlier period must be noted. This is the concept of a 'green belt', which was introduced in 1955. Green belts have been defined around many major cities, the most important being the metropolitan green belt around London. The purpose was to limit urban growth at the periphery of major conurbations by defining an area, mostly of farmland, woodland, or rural parkland, where existing land-uses would remain largely undisturbed. Urban growth which could not be accommodated within the cities was encouraged to take place beyond the green belt, possibly in new towns or areas of relatively high unemployment.

URBAN CONSERVATION

The scope of planning in relation to conservation of the built environment was significantly extended in 1967 by the Civic Amenities Act. Provision had existed beforehand, of course, for protection of individual buildings of importance, by including them on the secretary of state's list of buildings of special architectural or historic interest. Demolition or alteration of any such building requires a specific approval, although demolition does not otherwise require planning permission. However, this Act extended considerably the powers available to local authorities to protect their architectural heritage, principally by designating conservation areas, and by increasing the controls available within these areas. Emphasis was on the area within which individual buildings, or their grouping and layout, were of such a quality as to justify measures for their protection and enhancement. The concept of conservation rather than preservation was introduced. Conservation areas are not intended to be museum-pieces, but remain in normal use. Some new building is accepted, but it should be designed with particular care in order to contribute to the visual quality of the area being conserved. Measures to reduce traffic, pedestrianise streets and remove obtrusive features of the environment are often carried out by the planning authority as part of their conservation-area policies. Well over 3,000 conservation areas now exist, including many town or village centres.

REDEVELOPMENT

Another legacy of planning activities in the 1960s, but one which is much more alien to modern attitudes among planners and the public than conservation, is that of major comprehensive redevelopment and monolithic new developments in the town centres and residential areas. There have been major changes of attitudes to city-centre and residential development since then, and some of the least satisfactory high-rise housing developments have actually been demolished less than twenty years after their construction.

Apart from the visible presence of projects built during the 1960s or earlier 1970s, the significance of this period of planning in Britain lies in the responses and reactions which underlie much of modern planning practice. For instance, the trend towards more humane housing policies, for low-rise and traditional-style housing, improvement and gradual renewal, or the interest shown in social aspects of planning, and the links between planning and community work and social policy. These trends were linked to the incorporation of public participation in the planning process both by legislation and by widely accepted practice. Finally, the idea has been accepted that the inner city exhibits a distinct range of planning problems too complex to be tackled by traditional sectoral policies such as housing redevelopment.

The 1960s saw the redevelopment of housing, especially of the traditional nineteenth-century terrace housing, proceed at an ever-increasing pace. Schemes became more elaborate and ambitious both in terms of the numbers of dwellings cleared and of the design of the housing built in their place. As the impact of local authority housing projects became more apparent, criticism increased, chiefly on two grounds. The first was economic. The case was that redevelopment was a very costly way of providing dwellings of a good modern quality, and that many older dwellings could be improved to acceptable modern standards, giving them a thirty-year life, less expensively *pro rata* than the notional eighty-year life of newly built dwellings. The second main criticism was on social grounds. The bad social consequences of redevelopment became increasingly apparent. Architectural design concepts such as streets in the air, or vertical streets which, it was claimed, offered exciting new lifestyles, failed to live up to expectations. People clearly missed the scale of traditional residential areas, and the pleasure of having good kitchens, sanitation and hot water was insufficient compensation. New housing projects often lacked a good range of local shops and other community facilities. Several well-documented studies showed the adverse social and economic consequences for many households facing higher rent, and more expensive shopping, heating, or journeys to work. Planners generally recognised the need for shops, community facilities and adequate public transport, and provision for these was usually incorporated in plans for new housing projects. However, planning departments as such rarely had the power to ensure implementation of these facilities. Some, such as schools, presented no problem, but many housing schemes were less satisfactory not because the planners made inadequate allocation, but because they lacked the power to ensure the implementation of community centres, new bus routes, or provision of more than a few basic shops.

IMPROVEMENT OF HOUSING

Planners were not always among the most vociferous critics of the big housing projects, although many were quick to see the possibilities of the new approach to housing that came with the 1969 Housing Act, and the introduction of general improvement areas (GIAs). Improvement of dwellings to prolong their life or modernise them was not a new concept, but this legislation greatly simplified procedures, and at a time when there was a widespread desire for alternatives to redevelopment, authorities in many parts of the country took advantage of it. Designation of GIAs, in which environmental improvements of the area around the housing by means of pedestrianisation of streets, reduction of through-traffic, off-street car parking, play areas for children and landscaping were associated with the improvement of individual dwellings, became a very widespread policy.

Unfortunately, residential areas were frequently strictly categorised into redevelopment or improvement areas, and some areas of course were not entirely suitable for either. An additional area-based designation was introduced in 1974, the housing action area (HAA). This was an attempt to provide measures for the improvement of more marginal dwellings, but not necessarily to the full standard normally adopted in GIAs, while at the same time ensuring that the function of dwellings that catered for particular social groups with limited means, and limited access to the housing market, remained as they were.

Many of the housing programmes initiated in the 1960s continued well into the 1970s, but house-building is now on a much reduced scale. Newer concepts such as the gradual renewal of older housing areas are also being pursued. The whole picture is quite different from the programmes of the 1960s and early 1970s which were much more massive in scale, but simpler in conception than those of the present time.

PLANNING AND THE COMMUNITY

This greater complexity can be associated with a much closer relationship that has now been established between planning and social policy. The grandiose city centre, housing and new town designs of the 1960s were the more extravagant products of planning conceived primarily as a physical design activity. Of course, it has always been accepted that the planning was undertaken for the benefit of the people, but it was not until the early 1970s that the concept of planners working with the community for whom they were planning emerged strongly. The communication skills of the community worker, the social worker's perception of social problems and their causes, and the integration of planning policies with those of other agencies of social policy in the form of corporate planning, were eagerly sought as ways of improving the understanding and sensitivity of the planner. Public participation grew rapidly in importance for similar reasons.

Public participation was introduced in the Town and Country Planning Act 1968 in order to provide the means whereby planning decisions would have greater public acceptability. The duty to undertake public participation is quite explicit. Development plans cannot be approved without participation to an extent satisfactory to the secretary of state. Thus, public participation became central to statutory planning procedures, but it did not so readily gain acceptability in related procedures such as those for housing or road developments. These are both, however, subjects on which public confidence in decision-making and the outcome of policies is not as great as it should be. If public participation is justified as a means of generating public confidence, its incorporation in housing or transport planning procedures was necessary, but was only gradually achieved during the 1970s.

The advent of public participation in statutory planning procedures has caused planners to widen considerably the range of consultations with both individual citizens and organisations or interest groups. Presentation of planning proposals to the general public by means of exhibitions, leaflets, public meetings and other forms of publicity is an important aspect of the planners' work in preparing a structure or local plan. The issues about which people are concerned is frequently instructive. Often, of course, they are not planning matters.

It is difficult to say how much direct influence has been achieved by the general public as a result of public participation in planning. In general, organised groups have achieved more influence, and views of individuals have only been of real significance at a very localised scale. At the strategic level issues are often difficult for many members of the public to grasp.

INNER-CITY POLICY

Recognition that the inner city areas constitute a distinct planning problem dates from the late 1970s, although the research findings on which the idea is based, and implicit recognition of components of inner city problems, go back much earlier. In essence the problems associated with the inner city are loss of employment, loss of population to peripheral or new town housing, lack of new development or investment, and large areas of derelict or underused land or premises. The term inner city refers not to the city centre, but to the area around the centre, particularly those areas first developed in the nineteenth century in the major industrial cities. Some of the problems have followed from past planning policies such as migration of people and industry to new towns, redevelopment of housing areas and building overspill estates, and very importantly, the loss of many areas of small industrial premises.

An influential series of studies sponsored by the government, known as the Inner Area Studies, were undertaken during 1975-7. These studies by teams based in Liverpool, Birmingham and the London Borough of Lambeth[4] made an important contribution to identifying the nature of the problems, and possible policies. These studies were followed by a major change of policy by the Secretary of State for the Environment in 1976, when he announced measures to boost the resources available to the major industrial cities and improve their industrial base.[5]

Legislation was initiated, the Inner Urban Areas Act 1978, and an urban programme of direct assistance to local authorities with the greatest problems was established. Seven areas of high unemployment were designated Partnership Authorities. These were Newcastle/Gateshead, Manchester/Salford, Liverpool and Birmingham, and Islington/Hackney, Lambeth and Docklands in London. Extra resources were directed to these authorities by the government, and a special organisation, involving both central and local government representatives, was established to implement policies. Partnership money is being used for housing and environmental improvements, restoring areas of derelict land, assembling land for industrial development and a variety of other purposes.

One of the major innovations in planning in recent years, associated with inner city policy, is industrial promotion and the generation of employment. Local authorities have, since reorganisation, taken a major role in employment planning, as it is often described.

Of course, central government has for many years pursued policies designed to promote industrial or office development in the regions with higher rates of unemployment by offering incentives to manufacturers to move to designated regions, building advance factories and dispersing certain government offices. However, this used to be seen as something separate from the work of local planning authorities, and not closely linked with other aspects of planning policy. After 1974, the new authorities reviewed policies and priorities within their areas, and a number of authorities in the older industrialised conurbations identified employment promotion as a major priority. The measures adopted tended to be in the form of offering loans or grants to assist small industries, assembling land and building advance factories for industrial development and adapting the area-improvement concept to the task of improving older industrial areas.

Tyne and Wear County Council, and Newcastle City Council, in the north-east of England, were among the pioneers in employment planning. The approach they adopted was to obtain special legal powers by means of a locally applicable private Act of Parliament, the Tyne and Wear Act 1976. This Act gave the county and district authorities in Tyne and Wear powers to make loans or grants to industrialists in their area, and declare Industrial Improvement Areas (IIAs). The concept of an IIA was first promoted by another pioneering authority, Rochdale, near Manchester, and was based on an analogy with the housing GIA. An area of industrial land or old buildings in the inner area is designated as an area that will receive certain benefits. The risk of wholesale clearance is removed, thus generating confidence and encouraging investment. Loans and grants are made available, and environmental improvements such as clearing derelict land and eyesores, preparing sites for development and improving vehicle or pedestrian access are undertaken.

Experience in Tyne and Wear, and in Rochdale, served as a model for other authorities using later legislation that applies to all major cities, the Inner Urban Areas Act 1978. This Act provides similar powers, including the power to declare Improvement Areas. These are similar to IIAs, but may include areas of commercial as well as industrial land use. A number of these have now been declared in several major cities.

Factories built in advance of need are another aspect of employment planning. It is only in the late 1970s that local planning authorities have developed these to a significant extent. Previously only central government agencies undertook this. Local authority involvement has influenced provision in two ways. First, they locate their advance factories as the basis of assisting specific disadvantaged areas of the city, as is the case with other aspects of local employment planning. Secondly, local authorities have identified a gap in the range of central government advance-factory provision. Very small factories, and units large enough to be used only as small workshops, known as 'nursery units', have been most sought after, and local authorities are clearly fulfilling a real need. Behind much of this activity is the theory that new and small firms are the key to future prosperity and employment growth, which is accepted somewhat uncritically. The positive role of stimulating development, rather than simply controlling it, is one that planning authorities like to adopt, however.

Improvement Areas and the new concept of Urban Development Corporations seek to achieve their objectives by a highly interventionist approach as do many other aspects of planning policy in other sectors that have been reviewed here. The latest proposal by the government is for Enterprise Zones (EZs), which is exactly the reverse. The objective of economic development is sought by an absolute minimum of government regulation and intervention and the creation, as far as possible, of *laissez-faire* conditions.

Eleven zones have been designated, and came into effect in 1981. These range in size from just over 40 ha to around 400 ha, the biggest being the Newcastle/Gateshead zone. In 1982 the government announced that another eleven zones were to be designated, although it is still too early to judge how valuable the first ones have been in stimulating new development. Although not explicitly part of inner city or regional policy, they are mostly in areas with old, declining industry and above-average unemployment. Within the zone relief is granted from local taxation or rates, and other tax and financial benefits are offered. In addition, there is a simplified planning regime whereby permission to develop is automatically available within areas with the appropriate land-use allocation. The EZ concept represents a radical departure from orthodox British planning procedures. It is sometimes presented as an experiment in non-planning, but it should more correctly be regarded as an experiment in a different form of planning.[6] EZs will be judged by their success in stimulating development. Much of the development which takes place will be attributable to the favourable financial regime, and infrastructure investment taking place in many of the zones, rather than to the simplified planning regime.

CONCLUSIONS

As is clear from this review of British planning, the scope and range of its activities has broadened out greatly since the war. Very little has been lost apart, perhaps, from comprehensive redevelopment, but many themes, such as conservation, or corporate planning are taking a less prominent role now than earlier. Employment planning is prominent at the moment, but will also no doubt become assimilated into the system as new priorities emerge.

The emphasis throughout has been on the work of local planning authorities and other local planning agencies such as new-town development corporations, because planning in the UK is primarily a function of public administration. Professional planners are, of course, not the only professional group involved, but they play a leading role in policy-making and implementation in all the themes described, along with other groups such as engineers, surveyors, architects and landscape architects.

The role of the professional planner tends only to a limited extent to be the designer of land-use plans or detailed layouts. Much more of his time is spent collecting and analysing data in order to understand the changes in the economic or social or land-use structure of his area, negotiating the necessary details for implementation of projects, or analysing the impact of proposals for development for which planning permission is sought. The most important tasks tend to be the essential work involved in ensuring that proposals can be realised by co-ordinating the contributions of other departments, other authorities or agencies, or private organisations, and ensuring that sites and associated facilities or services are all available at the right time and that the necessary financial provision is made. Public participation also makes major demands on planners' time.

Planning does have the formal structure of a profession.[7] The term 'chartered town planner' may be used to describe only those people who are Members of the Royal Town Planning Institute, designated by the letters MRTPI after their name. The qualification is achieved by first passing examinations of the Institute or obtaining a degree or diploma from a university or polytechnic recognised by the RTPI for this purpose, followed by a minimum of two years' relevant practical experience.

Regulating professional qualifications by an institute run by its members and independent of government in this way protects the clients by ensuring that the professional service is offered by suitably qualified people only. In the case of planning the client is rarely an individual, and the majority of members (over 75 per cent) are employed by public authorities and, therefore, not paid by fees from individual commissions. The RTPI was founded in 1914, but it is only since planning became a major function of local government after 1947 that it became predominantly a public service profession. Traditionally a planning qualification was held in conjunction with architecture, or perhaps civil engineering or surveying. In the postwar period qualification in planning alone became more usual, with the first BA degree being offered at Kings College (now the university), Newcastle upon Tyne. In the last twenty years large numbers of people with degrees in economics, geography, social studies and other academic subjects have also qualified as planners. This variety within the profession allows planning authorities to employ people with a wide variety of expertise, who also share a professional qualification which enables them to contribute to the teamwork which is normally necessary.

Planning has been a growth sector of public administration since the 1947 Act, and especially in the last twenty years. It has been boosted not only by extensions in the scope of planning, but also by the effects of local government reorganisation in the mid-1970s, which split the planning function and created new planning jobs. Now local government has to economise, and the role of counties has been reduced. More than ever before, the planning profession needs to demonstrate its value to society.

NOTES: CHAPTER 8

1 See, for example, P. G. Richards, *The Reformed Local Government System*, 4th edn., London, Allen & Unwin, 1980.
2 This situation is analysed in A. Alexander, *The Politics of Local Government in the UK*, London, Longman, 1982, pp. 54–73.
3 Town and Country Planning Act 1971, section 22(1).
4 G. Shankland; P. Wilmot and D. Jordan, *Inner London: Policies for Dispersal and Balance: Final Report of the Lambeth Inner Area Study*, London: HMSO, 1977; H. Wilson, L. Womersley, R. Tym and J. MacKay, *Change on Decay: Final Report of the Liverpool Inner Area Study*, London: HMSO, 1977; Llewelyn-Davies, Weeks, Forestier-Walker and Bor, *Unequal City: Final Report of the Birmingham Inner Area Study*, London: HMSO, 1977.
5 This change is discussed in McKay and Cox, 1979, pp. 251–6; and Lawless, 1981, pp. 7–13.
6 This argument is developed in R. H. Williams and P. Butler, 'Enterprise zones; dogma abandoned', *Town and Country Planning*, vol. 51, no. 3 (March 1982), pp. 69–71.
7 A history of the profession is found in Cherry, 1974b.

Chapter 9

Ireland

K. I. Nowlan

Planning in Ireland is based on laws that are deceptively similar to those of England and Wales, but there are in fact a number of distinctive features. Modern attempts to establish a statutory planning system to regulate development date from 1934, when the Town and Regional Planning Act was passed. Local authorities at that time displayed little interest in planning, and their own development, especially of housing, demonstrated indifference to principles of good planning. The Act of 1934, and the Amending Act 1939, applied to less than half the area of the country, since they needed to be specifically adopted by individual local authorities before coming into operation.

Though physical planning was not popular among local councillors, the requirements of local government legislation and the administrative control of building exercised by the Department of Local Government, now the Department of the Environment, ensured that individual public developments were reasonably well built, even if built in the wrong places. The 'jungle' of local government law, as an eminent judge described it, is made up of dozens of laws and parts of laws enacted during the nineteenth and early twentieth century. For instance, the Housing Act 1966 repealed a total of forty-eight Acts of Parliament. Similar spring-cleaning has not been done in other areas of local authority law. Consequently, although physical planning law is modern and generally effective, it cannot be operated in isolation from other laws relating to buildings, roads and sanitation. In particular, the Housing Act 1969 overlaps the planning Acts in a number of important matters. The Roads and Motorways Act 1974 also has impact on the basic planning law.

The principal Act providing for the control of physical development is the Local Government (Planning and Development) Act 1963. It is amended by the Local Government (Planning and Development) Act 1976. The objectives and scope of the planning system that was envisaged by those responsible for this legislation is best summed up by

104   *Planning in Europe*

quoting from the speech of the Minister for Local Government, when introducing the Bill to the Senate in 1963.

> The purpose of the Bill is to provide the legislative framework for the comprehensive, orderly, and progressive planning of cities, towns and rural areas. The Bill is intended to provide a more workable and flexible planning system to be operated by planning authorities. The system will enable planning authorities to exercise comprehensive control on development in their areas and, what is more important, to engage in positive development and redevelopment work themselves. It will enable them to facilitate industrial and commercial development and to secure the redevelopment of those parts of built-up areas which have become outmoded, uneconomic or congested; it will also confer wide powers aimed at securing the preservation and improvement of amenities in town and countryside.
>
> Planning authorities will be required to make development plans for their areas within a period of three years. Each development plan will consist of a written statement of development objectives and a plan to illustrate these objectives. Certain minima are prescribed. In the case of cities and towns, these are: zoning of land uses for particular purposes, whether residential, commercial, industrial, agricultural or otherwise; securing the greater convenience and safety of road users and pedestrians by the provision of parking places or road improvements or otherwise; development and renewal of obsolete areas; and preserving and improving amenities.
>
> It is my aim that, following on the passage of this Bill, regional studies should be made with strong support and guidance from my Department in order that the economic objectives of planning that I have outlined will be pressed forward vigorously. Other countries in Europe are using physical planning to advance their economic development and we cannot afford to ignore their example. Indeed, I regard this as probably the most important long-term task facing the local authorities in this country. Planning for growth seeks to develop dynamic centres which will have the economic strength to prosper and to support the public services, the entertainments, the amenities and the shops which people expect nowadays.
>
> As regards national planning, I am fully in favour of the maximum possible degree of leadership, advice and assistance being given by the central authority. It is my intention to have my Department make a much greater positive contribution to planning in this country than has been possible under the Act of 1934.[1]

In the particular parts of the ministerial speech quoted, only passing reference is made to environmental quality, but the Act does not neglect this important aspect; and the duty of preserving and improving amenities is imposed on planning authorities as a matter of the utmost importance. The operation of the planning Acts and all building control is vested in local authorities. These authorities, both urban and rural, are independent of central government in their planning decisions, though their independence in many other matters is considerably modified by their heavy reliance for finance on central government, and in certain specific circumstances the Minister for the Environment may issue a direction to local planning authorities.

Have the hopes expressed in the minister's speech been realised? The response to such a question must depend on the standard of action demanded. Every planning authority has made a development plan. Many have made the statutory review postulated by the Act. Some authorities made real efforts. They have employed qualified planners and produced useful results. The majority of authorities have complied with the mimimum requirements of the Act. Others have still to learn that 'proper planning and development' cannot be achieved in the absence of competent planners.

THE LOCAL GOVERNMENT SYSTEM

The Irish local government structure has been little changed since Ireland achieved independence in 1922. The principal structural change was brought about by the removal of the lower tier in county areas. Only in relation to towns within counties is there any semblance of two-tier administration.

The state is divided into twenty-seven counties and four county boroughs. The latter are the four largest towns of Dublin, Cork, Limerick and Galway. They in themselves have county as well as urban status and powers. The remainder of the state is divided into twenty-seven administrative counties. Within most counties there are towns with the status of borough or urban district, which have the same powers and functions as each other. Lower still in the administrative hierarchy are small towns over which town commissioners exercise very limited powers. There are eighty-seven planning authorities, namely, the county boroughs, boroughs, urban district councils and county councils.

Though the structure of local government is so little changed from that created at the end of the nineteenth century under British rule, the internal administration of the actual authorities has been fundamentally revised. It bears little resemblance to that which existed fifty years ago. Much of the change follows from legislation which has added to or transferred away some of their functions, but the most striking changes have followed from the creation of the entire local government service as a single service for employment purposes and the development of the system of county and city management.

The management system was first introduced by separate statutes into the cities of Cork and Dublin. Later it was extended to Limerick and Waterford and, finally, in 1940-2 to all the remaining local authorities. Its basis rests on a separation of all local authority functions into reserved and executive functions. For the purposes of the former, the elected council acts by resolution; for the latter, the manager exercises the powers of the authority by order. Every function is an executive function unless statute or statutory order defines a function as reserved. Under planning legislation the making of a development plan and of its variations is a reserved function. The granting of permission to carry out development is an executive function. Very broadly, policy is determined by reserved function, operations are controlled by executive order of the manager. The manager is a permanent full-time officer of his authority. In each administrative county the manager is manager for each borough or urban district.

THE PLANNING SYSTEM

The operation of the planning code is based on three sections of the 1963 Act. Each section has been amended to a minor degree by the Local Government (Planning and Development) Act 1977. This latter Act will be discussed later, but where particular sections of the 1963 Act are discussed, it is the amended versions that are referred to.

The first basic sections of the 1963 Act prohibit building or construction work or material change of use without permission of the planning authority, and provide for grant (or refusal) of permission to carry out development. A duty is imposed on each planning authority to make a development plan with certain minimum content. There is an extensive list of objectives set out in the Third Schedule to the Act, which may be included in the development plan if the authority so chooses.

The concept running through the two Acts is that of controlling development which it defines as 'the carrying out of any works on, in or under land or the making of any material change in the use of any structures or other land'.[2] The protection of environmental amenity is a predominant objective. Planning authorities are themselves prohibited from carrying out development which would conflict with the development plans which they have made.

THE DEVELOPMENT PLAN

The 1963 Act came into operation on 1 October 1964. Each planning authority was required to make a development plan within three years, and to review it at least once in every five years. Plans have been made in most areas and reviewed at least once. For every urban area, the development plan must contain the minimum elements, as indicated in the minister's speech, and in addition such optional matters listed in the Third Schedule as may be desired. For rural areas, a minimum content is similarly prescribed together with an optional content selected from the Third Schedule, plus an optional objective of use-zoning.

The development plan is adopted by formal resolution of the planning authority for the area, and it is not subject to approval by any other authority. The Minister for the Environment, however, is empowered to require that the plans of contiguous authorities be co-ordinated, and he has in addition a general power to require a planning authority to vary its plan in such manner as he may direct. The development plan is binding on the authority which made it.

Before a development plan can be adopted, a draft of the proposed plan must have been exhibited for a period of three months. Certain persons who are particularly affected by its provisions must be notified. Anybody may object in writing to any or all of its provisions. Any ratepayer may demand that his objection be heard by a person appointed to hear objections. The planning authority must take all objections into consideration when determining the content of the development plan.

Though the statutory requirement for public participation is not onerous, the influence of pressure groups of one kind and another, including that of the National Trust, An Taisce, has been sufficient to delay inordinately the statutory review of some development plans.

With few exceptions the development plan is the controlling instrument for development in its area. The local authority is specifically prohibited from carrying out development which would conflict with it. No authority except departments of state headed by a member of government is exempt from planning control. The planning board in considering appeals (see below) must have regard to the development plan, but is not bound by it.

The development plan consists of a written statement and a plan (maps and diagrams), indicating the objectives for the area. There is no requirement that any maps be to scale or that they indicate precise details of objectives. The more sophisticated development plans include statements of policies which are intended to be implemented through the chosen objectives selected from those set out in the Act.

The development plan has three major purposes. The first is to require the planning authority to co-ordinate its policies and to plan for them. The second purpose is to ensure that the public is made aware of authorities' plans and is given an opportunity to influence them. The third purpose is to create a background against which the propriety of proposals for development can be examined. The zoning objectives in urban areas can have considerable effect on the value of land and buildings.

CONTROL OF DEVELOPMENT

The definition of development was given above. In principle all development requires permission, but certain types of development are defined as 'exempted development' and some forms of development may be exempted development in particular circumstances. Development is exempted primarily for the practical purposes of permitting everyday work to be carried out without unnecessary interference by the authority and to reduce the burden of development control on authorities. For example, almost all structural work to the inside of normal buildings is exempted development. Most agricultural work is exempted. Work which is normally exempt may not be exempted in an area to which a Special Amenity Area Order applies. It is a criminal offence to carry out development, other than exempted development, without permission either of the local planning authority or of the Planning Board on appeal.

Permission to carry out development must be sought by applying to the planning authority for the county or urban area in which the site of the proposed development is located. Particulars of all applications and decisions on them must be entered in a register which is available for public inspection. The actual application is also available for inspection. When authorisation is sought for a private industrial project, the local authority may require the developer to furnish an impact study with the application, if it feels that the environmental consequences may be significant. This provision does not apply to public projects.

The planning authority must make a decision and serve notice of its decision on each application within two months, unless the applicant has agreed to a time extension. Failure to serve notice of its decision within the time limit is deemed to be a decision to grant permission free of any condition. Appeal may be made against such decision of the planning authority to the Planning Board (see below), and the applicant for planning permission may lodge his appeal within one month of notification of the decision. Any third party may lodge an appeal within twenty-one days of the decision.

The planning authority for the purposes of granting or refusing permission is the county or city manager. In deciding whether or not to refuse permission or to grant it with or without conditions the manager is required to consider the proper planning and development of the area of the authority, the preservation and improvement of its amenities, and the provisions of the development plan, or of any special amenity-area order relating to the area. Particular conditions which may be attached to a permission include the payment of contributions to the cost of infrastructure to be provided by the local authority.

The Housing Act 1969 impinges upon the operation of development control by restraining a planning authority from granting permission for any development which involves the destruction or change of use of a dwelling unless a consent under that Act has previously been obtained from the housing authority or from the Minister of the Environment after appeal. Responsibility for housing is held by the same local authorities as planning, and the city or county manager is responsible for executing housing as well as planning powers. If the circumstances justify it, permission to carry out development in conflict with the development plan may be granted with the consent of the elected council and after public notice and consideration of any objections that may be made.

Every development control decision of a planning authority may be the subject of an appeal to the Planning Board (An Bord Pleanala). The right to appeal is universal, provided that it is exercised within the specified time. The fact that public notice must be given of all applications to the planning authority means that any person whose interests might be affected by the development proposed is forewarned and consequently the number of appeals is considerable.

To regularise developments which have been carried out without planning permission applications may be made for the retention of structures or use changes which have been made without planning permission. In such cases the planning authority is required to exercise the same kind of judicial consideration as in the case of an application for permission. There is a right of appeal against the decision of the planning authority.

THE PLANNING BOARD

Under the provisions of the 1963 Act all appeals against decisions of the planning authority were made to and determined by the Minister for Local Government. Because of his other functions under the Act, certain anomalies arose. Complaints were made that from time to time ministers had allowed party politics to influence their decisions. The administrative burden of determining 3,000–4,000 appeals per year was real and, in some instances, very long delays occurred. To remedy these ills the Planning Board was created by section 3 of the 1976 Act.

The Planning Board is free and independent in making its decisions. It must state reasons for its decisions, and it must note the policies of public authorities, the activities of which have a bearing on proper planning and development, and on amenities.

The Board consists of a chairman who is appointed by the government, and between four and ten other members who are appointed by the Minister for the Environment. The persons appointed are required to devote their whole time to their task. The chairman must be a former high court or supreme court judge. Board members, other than the chairman, are appointed for three years. The Board as first constituted consists of a former high court judge as chairman, two civil servants on secondment, of whom one is a professional planner, one county council officer, one trades union official and one woman architect.

Appeals to the Board may be heard either by way of public oral hearing or written submission. A party to an appeal, that is, the appellant, applicant, or the local planning authority, may request an oral hearing. The Board may agree or refuse to allow an oral hearing. In the event of a refusal to allow an oral hearing there is a right of appeal to the Minister for the Environment to direct the Board to hold an oral hearing. During 1978 the Board determined 2,487 appeals; it held 390 oral hearings. The decisions of the planning authorities were reversed in 25 per cent of all the appeals determined.[3]

When considering an appeal the Board is required, in the first instance, to have regard to the same constraints as the planning authority from which the appeal derives. It is not bound by the development plan of the authority and it must take into account the policies of the Minister for the Environment and of any public authority whose functions have a bearing on planning and the preservation of amenities. The Minister for the Environment may give general directions about policies but may not exercise any power in relation to a particular case. Every directive must be published to each house of the legislature and notified to every planning authority.

The Board may request an applicant to submit a revised proposal in lieu of the original application. Before arriving at its decision on a case, the Board receives the advice of an inspector who is a person qualified or experienced in planning matters. The decision of the Board is not subject to further appeal. Although the Board or any of its members may personally preside at oral hearings, it has preferred to delegate this function to inspectors. Oral hearings are usually held in or near to the office of the authority which has made the decision against which the appeal is made.

The Board is required to state its reasons for refusal or grant of permission, or conditions which it may attach to a grant of permission. It is not required to publish the report of the inspector who has made a recommendation. In the case of land being incapable of reasonably beneficial use, when planning permission has been refused after an appeal, or onerous conditions attached, a purchase notice may be served on the planning authority requiring it to purchase the land.

The Planning Board is a distinctive feature of the Irish planning system. It is an innovation that has no parallel in the British system.

ENFORCEMENT OF DEVELOPMENT CONTROL

In 1963 provision was made for the imposition of penalties for carrying out development without permission, but the penalties were small, and have not been increased. In areas where the unauthorised development proves to be harmful from the planning point of view an enforcement notice may be served, requiring the alteration or removal of the structure and restoration of the land to its former condition. An enforcement notice may require the discontinuance of a use which has been established without permission. Failure to comply with an enforcement notice is an offence punishable by fine. The planning authority may enter and take steps to remove an unauthorised structure and restore land, and the cost of such action may be recovered from the owner of the land.

Both the procedures of the district court and the penalties imposed in cases where prosecutions were successful have been found by planning authorities to be unsatisfactory. Some planning authorities are alleged to have abandoned district court actions in planning cases. The Planning Act 1976 provided new procedures and removed some of the constraints on district court actions. It extended the period during which prosecutions might be taken from six months to five years, and introduced two new forms of action. The more important of these enabled any person, including the planning authority, to seek an order from the high court directing the immediate cessation of unauthorised development or the discontinuance of an unauthorised use. Numerous petitions for orders have been successful. The high court has wide powers to secure compliance with its orders. At least one person has been imprisoned for failing to obey.

The second new enforcement procedure in the 1976 Act is that of serving a 'warning notice'. Failure to comply with a warning notice may lead to fines of £250, or in a case of continuing defiance of a notice, to a daily fine of £100 and six months' imprisonment.

All the notices mentioned above relate in one way or another to unlawful development. The 1963 Act also gives power to the planning authority to require the removal or alteration of a building or land use which enjoys the benefit of complete legality. There is a right of appeal to the Planning Board against the provisions of such an order. Compensation is payable for damage suffered as a consequence, unless conditions are imposed with the object of avoiding or reducing serious air or water pollution, or with that of stopping the exhibition of advertisements when compensation may not be payable.

DEVELOPMENT BY THE PLANNING AUTHORITY

One objective of the 1963 Act was to encourage local authorities to promote development actively. It was intended to give planning authorities the power to assist private developers to acquire parcels of land necessary for satisfactory development. An additional power enables the authority to enter into partnership with 'any person or body for the development or management of land'.

In connection with their housing activities local authorities have provided sites for factories, shops, schools, churches, and so on. At least one authority has acquired and cleared land in a run-down city area and secured its redevelopment through a development company. The arrangement is entirely financial. It is unlikely that a business concern could make a workable co-management agreement with a local authority. There is no report of a planning authority using the power given to it to assist a developer in the acquisition of land.

CONSERVATION AND AMENITY

Although almost every decision which a planning authority is empowered or required to make is supposed to be related or based upon the preservation and improvement of the amenities of its area, these basic matters tend to become overlain by a blanket of other requirements. The preamble to the 1963 Act refers to provision for the preservation and improvement of amenities. A planning authority, when considering applications for permission for development, can ensure that proper open space is provided and that new buildings conform in character, size, height and material to acceptable standards. Conditions to secure these standards may be imposed.

The cost of complying with such conditions must be borne by the developer. The authority can also make provision for the reservation of land as game or bird sanctuaries, as public open space, or for the preservation of objects of archaeological, geological, or historic interest. Views and property and places of natural beauty may be protected, and a planning authority can list for preservation features of artistic or other interest forming the interior of structures.

An area may be declared to be of special amenity because of its outstanding natural beauty, its special recreational value, or a need for nature conservation. The declaration may limit the kinds of development which may be permitted in the area. The making of an order declaring an area to be of special amenity in the first instance is a reserved function. That is, it can be made only by resolution of the council for the area.

The second of the three important amenity powers is that of making conservation orders. These may be made after consultation with specified conservation bodies to protect specific flora or fauna. Such an order may be made only in an area to which a special amenity-area order applies. The order is made by the city or county manager. A person may appeal to the Planning Board against such an order.

Tree preservation orders may also be made under the 1963 Act. Such orders may prohibit absolutely the felling of particular trees and woods or prohibit felling of trees without the specific permission of the planning authority. There is a right of appeal to the Planning Board.

PERSONNEL

An innovation in local government law is contained in the 1976 Act, which requires members of local authorities to enter in a public register statements of interests in land in the county or town for which they are elected. Certain employees of planning authorities and employees of the Planning Board must make similar registered declarations. Members of the Board must declare their interests in land held anywhere in Ireland.

The planning profession in Ireland is subject to strong British influences. Many have received their planning education in Britain or Northern Ireland, although there is a well-established Department of Regional and Urban Planning at University College, Dublin. A professional institute of planners in Ireland was founded a few years ago, but many planners in Ireland are members of the Royal Town Planning Institute, and a local branch of the RTPI operates in Ireland. Local planning authorities rarely employ large numbers of planners, and their work tends to be concentrated on development control rather than development plan making or general policy issues.

NOTES: CHAPTER 9

1 *Parliamentary Debates*, Dail Eireann, Dublin, Stationery Office, 31 July 1963.
2 Local Government (Planning and Development) Act 1963, s. 3.
3 *Annual Report* (Planning Board), Dublin, An Bord Pleanala, 1978.

Chapter 10

Denmark

O. Kerndal-Hansen

Danish planning legislation has been changed radically during the 1970s. A general land-use Act in 1970 was followed by an Act for national and regional planning in 1973, and an Act for planning at the municipal level in 1977, which provided local authorities with more refined planning instruments. During the same period major changes have taken place in society which are very important for planning. The economic boom in the 1960s and the early 1970s enabled many families to improve their housing standard and obtain a private car. The main theme of planning during the period of economic growth was housing construction, with priority given to single-family homes, followed by public investment in the necessary infrastructure, shopping and facilities like schools, kindergartens and nursery homes. A feature of this period was that planning failures in the economic field could often be overlooked in the context of a high rate of economic growth combined with high inflation.

Economic recession following the energy crisis of 1973–4 changed the working environment for planners considerably. The context now is of unemployment, balance of payments deficit and the prospect of a reduction in real income.

For planning, this situation has meant a change from urban development to urban renewal or improvement. New residential areas must still be planned, but development tends to be on a smaller scale. Housing construction has fallen to around half the previous rate, and housing projects are no longer on a massive scale. Now the great challenge for planning is urban renewal. In the older urban areas about 400,000 dwellings, corresponding to 20 per cent of the total number of dwellings in Denmark, are inadequate for reasons of size, internal amenity, or environmental quality. There has for some years been legislation for slum clearance, and Parliament (*Folketinget*) has recently considered a special Act to promote urban renewal with the aid of substantial subsidies.

In view of the recent reforms of Danish planning legislation, the consequences of which cannot be clearly predicted at present, and since urban renewal is now gaining importance, this account deals mainly with the goals and the principles of the new planning Acts and how the various types of plans are brought about.

EARLY TOWN AND COUNTRY PLANNING

As in other Western European countries, Denmark experienced extensive urban development in the years after the Second World War with a tendency towards increasing urban sprawl. Industry moved out from cramped inner-city locations to spacious industrial areas in the suburbs, as residential suburbs of family homes developed rapidly. Consequently, the demand for land was very great.

In this expanding economy there was a great need for an effective planning system. However, planning Acts were few and insufficient to control the development. There were no laws governing national or regional planning, and at the municipal level there were laws regulating urban areas but no obligation for the municipalities to plan for the areas of open country. Another characteristic was that planning was strongly controlled by central government agencies, which had the authority to approve and reject all town plans, even by-laws for small local areas.

Although no law on national planning had been passed, the government established in 1961 a national planning committee in order to establish strategies for public investment to guide location of industry and future urbanisation. In 1962 the committee submitted some recommendations in the form of a draft zoning plan in which the total area of the country was divided into four zones depending upon the present and future use of the area. The draft plan was intended as a basis for the local planning in the municipalities. However, the 'zone plan' was never politically discussed or approved by Parliament. After 1962, until the committee was dissolved in the mid-1970s, the main work of the committee and its secretariat was a range of analyses and forecasts concerning population, housing, industry, traffic, and so on, but no further plans were submitted.

For many years the greatest development took place in the Copenhagen metropolitan region and around the biggest provincial towns. However, successive governments sought to distribute the growth of population and employment over the country, so that in all regions it would be possible for the population to attain a share in the general economic, social and cultural progress of the country. In 1958 a law on regional development was passed providing government subsidies for industrial investment in peripheral regions. Since 1970 there has been a shift in the demographic trends in Denmark. Due to a lower birth rate, the population has grown at a slower pace and the population in the metropolitan area has tended to stagnate. Such population increase as there has been took place in the provincial areas.

Regional development subsidies and the overall tendency for the industry to move away from the big towns has, no doubt, had a great influence on the shift in the pattern of population growth. But the big expansion in the public sector which created many new jobs, especially at local government level, was also very important.

Prior to the 1970s there was no system of national or regional physical planning, although under the 1949 law for urban development some of the larger municipalities co-operated with surrounding smaller municipalities to prepare zoning plans for urban development covering fifteen years. However, the plans could not be called regional even though some of the biggest towns extended their plans over a large area. The only exception was in the metropolitan region of Copenhagen, where in 1947 the so-called 'fingerplan' was prepared.

The principle of this plan was that the urban development should take place along five suburban railway lines ('fingers') stretching out from the inner city of Copenhagen which formed the palm of the hand. The plan was based to a great extent on public transport with community centres placed at the railway stations, local centres in the residential areas and green wedges between the fingers. With the ever-increasing number of private cars and a growing proportion of single-family houses, the original assumptions of the plan became invalid, and a revision of the plan was made in 1960. Following this revision the government appointed a commission to work out a development plan for one of the fingers, along the bay of Køge. This finger was the least developed and the plan is the closest Denmark has so far got to a new town. In 1967 planning work for the metropolitan region resumed, resulting in a regional plan being approved in 1979 under a new law on regional planning in the metropolitan region.

In the municipal planning sphere the Town Planning Act 1938 was the main instrument until the new Municipal Planning Act 1977. Under the 1938 Act two types of plans were available to the municipalities, a structure plan (*dispositionsplan*) and a town planning by-law (*partiel byplanvedtaegt*).

A structure plan indicates the future aims for an urban area, and the policies designed to achieve the goals set. In fact, the structure plan was not prescribed in the Town Planning Act, and originated when the Ministry of the Environment in a circular letter in 1939 advised the municipalities to start their planning work by a structure plan.

The town planning by-law was usually worked out for a smaller area such as a precinct in a city centre. It specified details of land use, design, parking spaces, and so on. In the bigger towns the town planning by-law had to be in accordance with a building by-law which contained general rules for plot ratios in various parts of the urban area. Land-use details might be very precise. There are examples forbidding certain uses within department stores, for instance. Unlike the structure plan, the town planning by-law was binding for the propertyowners as well as the municipality. The by-law had to be approved by the Ministry of the Environment – which in some cases required a structure plan to be worked out before it would grant approval. Although the Town Planning Act was passed in 1938, it was not until the 1960s that it became common for the municipalities to prepare a structure plan.

A feature of the earlier municipal town plans was that they were physical plans and only covered the urban areas. There were no plans for rural districts. In most cases there were no regional guidelines within which the structure plan could be prepared. Furthermore, the structure plan was not a co-ordinated physical and economic plan. Consequently, municipal structure plans frequently proved to be unrealistic, grossly overestimating the regional importance of the municipality with regard to population growth and employment. Moreover, they did not take into consideration the economic resources necessary to implement the physical development indicated in the plan.

PLANNING REFORMS

The need for efficient planning instruments is very great in a rapidly changing society, and in 1963 proposals to reform Denmark's land legislation, seeking to create the necessary relationship between urban development, development in the open countryside and the need to keep down the rapidly increasing prices of land, were submitted to Parliament. After a fierce public debate, the whole complex of laws was, however, rejected by referendum.

Apparently attitudes changed quickly because only six years later, in 1969, a Liberal government was able to rally support for the passage of two Acts, the Urban and Rural Zones Act and the Nature Conservation Act, which were in some respects more restrictive than the Bills rejected in 1963. These two Acts formed phase one of a radical reform of planning legislation. As mentioned earlier, phase two was the Act on national and regional planning in 1973 and phase three the Act on municipal planning in 1977.

These reforms represent a fundamental change of principles underlying planning legislation. Goals, methods, procedures and the substantive scope of planning were all changed. In addition, the attitude to public participation in planning has altered, and it is now acknowledged that a framework must be agreed upon by the parties concerned before a final solution to any planning problem may be reached. In other words, the political nature of planning is more explicitly acknowledged.

Since the various planning provisions were incorporated in a number of laws, there was a need to simplify and modernise the planning legislation. The intention was also to establish a coherent planning system, ensuring that planning decisions were made on the basis of overall considerations of impact on society and that planning on the national, regional and local level was co-ordinated.

Two very important objectives of the planning reforms were the decentralisation of planning decisions and an increased public influence on planning. During the late 1960s and the 1970s there was strong pressure to allow decisions to be taken as close as possible to the people affected. A tendency to decentralise decision-making has, consequently, been seen in educational institutions, in industry and many other sectors of society, as well as in planning. Normal practice in physical planning had been for all formal plans, such as structure plans and local plans, to be approved by the central government, as were municipal decisions in other sectors. Admittedly, central government funded most municipal expenditure, but municipal authorities wanted greater freedom to make their own decisions, and planning reforms were initiated in order to comply with this wish. It came to be accepted that decisions should as far as possible be made by the authority directly responsible for tackling local problems, and that central government should be relieved of making detailed decisions on individual cases in order to minimise central control of municipalities.

At one time there were about 1,400 municipalities, but in order to form more appropriate functional units it was decided to reorganise the entire system of local government. The result was the formation in 1970 of 275 new municipalities. At the same time the number of counties was reduced from 24 to 14. The county councils had previously had a chairman representing central government, but as a result of this administrative reform, county councils now consist of local politicians only who elect their chairman, or county mayor, from among themselves.

Alongside increased municipal autonomy a change in the financial support system was also made. Previously central government refunded fixed percentages of municipal expenditure, but the municipalities were now to be given a lump sum. The lump sum was decided by such objective criteria as population, number of schoolchildren, aged persons and the total mileage of the public roads in the municipality.

The provisions for decentralisation contained in the new planning laws give the municipalities power to approve their own structure plans and local plans in so far as they are in accordance with the regional plan. The county council which works out the regional plan must, however, submit the proposed regional plan for approval by the Ministry of the

Environment. The new planning laws also give citizens, firms and organisations the opportunity to become involved in the planning process. Besides the general objective of creating the basis for more democratic planning, this is designed to counteract the increased remoteness of citizens from local authorities as a result of the larger municipalities formed in 1970.

Until the early 1970s there was no opportunity for public involvement in plan preparation until after a plan had been agreed by a local authority, prior to submission to central government for approval. This procedure gave the ordinary citizen little or no opportunity to influence planning decisions. In the new laws the process of planning has been phased to permit citizen participation at an early stage. The planning authorities come up with the first move, but the citizens are given an opportunity to voice their opinion before a final plan is proposed.

The new laws give citizens a number of rights. They must have easy access to information about planning. At certain stages in preparing a plan the planning authorities are required to publish analyses and other material and to stimulate public debate on planning issues. The law stipulates that propertyowners, residents and others within the area to be included in a proposed plan must be given written notification of its provisions, and that people have the legal right to make objections and representations. Power to decide the content of plans remains with the municipal councils, and there is no guarantee that the council will comply with the views put forward by citizens individually or in groups. There is no doubt however, that citizens' views will have great impact on plan content in the future.

Citizen participation in planning takes various forms. The most straightforward is public information, or one-way communication, from the planning authorities to the citizen, by advertisements, press releases, reports and exhibitions. Another is public discussion, or two-way communication, on the initiative of the planning authority by means of public meetings and local conferences. At a higher level of participation comes direct co-operation with citizens, with opportunities to contribute ideas in seminars or group studies. Local referendums on road or shopping-centre proposals may also be held. Finally, there are examples of self-determination, when the preparation, but not necessarily approval, of a policy to overcome a planning problem is delegated by the municipal council to the citizens affected by it. A typical case is that of an association of houseowners preparing a local plan for their own residential neighbourhood.

With citizen participation now mandatory in various stages of the planning process, it is claimed that planning procedures take too long. However, people increasingly want to influence the development of their own neighbourhood and general conditions of life.

MAIN PRINCIPLES OF PLANNING LEGISLATION

The high degree of decentralisation introduced into town and country planning freed the central administration from decisions in specific cases, but controlling guidelines were introduced by central government for planning decisions in the counties and municipalities. One of the main arguments in favour of delegating planning powers is that the county and municipal councils are capable of making decisions across the normal divisions of public administration into the educational sector, the social sector, and so on. Elected councils are especially fitted to consider issues that may not be compared directly but which have a political character, such as whether a new by-pass be constructed or the money used for a new swimming-bath.

The autonomy of county and municipal councils is, however, limited by the general guidelines which were set up by the central administration for the various sectors, and which must be followed by the county and municipal councils. These guidelines create a framework for decentralisation while allowing central government to continue to formulate national objectives within each sector and co-ordinate different sectors.

The National and Regional Planning Act stipulates that, in the first phase of the preparation of a regional plan, the municipal councils submits proposals to the county councils indicating the goals which the municipal councils wish to see pursued by the regional council. With the first planning proposals coming from the lowest level in the formal planning hierarchy, this procedure is therefore different in principle from traditional regional planning. The more orthodox procedure, following practice in other European counties, would be for an overall strategy first to be prepared as the basis for regional planning. When the regional plan has been worked out, the municipal planning authorities have the regional framework within which the municipal structure plan may be formed. This, in turn, forms the basis for local plans.

The idea of starting the planning process by proposals from the bottom up does not mean that the ultimate plan will be formed as a sum of the wishes put forward by a lower planning level. Its main purpose is to decentralise planning. Of course, in practice, views and decisions on strategic projects, such as motorways, bridges and universities on the national level, and hospitals and roads on the regional level, will form part of the basis upon which local planning takes place. The new element in planning is that the objectives are not decided once and for all by a superior planning authority, but formulated instead through interaction between strategic and local interests.

Another new element in Danish planning legislation is the rule that before the final proposals for a regional plan are worked out, the regional council must publish alternative draft proposals and state the assumptions upon which the drafts are based. The alternative drafts are phase two of the regional planning process. The alternatives, which should deal with such vital subjects as the urban structure, town and district centres, location of major public institutions, and so on, must be wide in their scope, so that the alternatives really cover different proposals. In the official guidance from the ministry it is even suggested that it may be appropriate to work out unrealistic alternatives. If such alternatives are rejected in the end and not published, the reasons should be given in the statement presenting the alternatives.

The Municipal Planning Act also aims at participation by citizens, firms and organisations in formulating the broad guidelines for the municipal structure plan. At the start of the planning process the council must publish a brief account of the main planning problems and possibilities, and stimulate a public debate, giving the public an opportunity to voice their opinions before a plan proposal is worked out. Alternative drafts are not required before a proposal for a municipal structure plan is prepared. To some extent, however, the report on planning problems and possibilities referred to above performs the function of a statement of alternatives. After public debate, the municipal council may adopt a proposed structure plan. In connection with the ensuing publication of the proposal the municipal council must also publish any diverging views on the proposal held by a minority in the municipal council.

With a great degree of decentralisation in planning decisions and public involvement in the planning process, it is now common practice to illustrate the consequences of plans for various interest groups. A plan that may be favoured by one group of the population may be unacceptable to another group, such as car owners and non-car owners, or private developers and public authorities. The planning laws do not explicitly provide for such descriptions of the consequences, but the official instruction from the Ministry of the Environment recommends that the alternative drafts for a regional plan are accompanied by a description of how the various drafts affect the interests of different groups of citizens.

Traditionally physical plans were not harmonised with economic plans, with the result that many physical plans were not implemented because of a lack of economic resources. Therefore, politicians and planners both wanted the new planning laws to make integration of physical with economic and even social planning possible. As it turned out the new planning laws did not contain strict rules for integrated physical and economic planning. However, the Act for national and regional planning provides that long-term economic objectives of the regional and local authorities must be observed when preparing a regional plan. The municipal planning Act requires the relationship between the structure plan and economic plans of the municipality, and planned investment in urban development and urban renewal to be described. Although planning legislation does not require integrated planning, many Danish municipalities have in the last few years found it appropriate to integrate physical, economic and social factors in a long-term plan.

In earlier planning legislation it was only possible to purchase property compulsorily in order to implement a plan if the purpose was to establish public facilities like schools, roads, and so on. Now, under the Municipal Planning Act, the municipal council is able to expropriate property when it is essential to the implementation of a local plan. That means that expropriation can also take place if the development is purely private, such as a housing project or a shopping centre, as well as for municipal projects. In addition, at the time of the first phase of the planning reform another Act was passed which provided for the municipality to have first option to purchase any property over 6,000 square metres that came up for sale. However, the fundamentally restrictive and regulatory principles embodied in the planning system remain after the reforms. Private persons, firms, or organisations can still not be ordered by the planning authorities to act in a specific way to implement a plan. (An exception to this rule is, however, the Slum Clearance Act 1969.) The plans can only set up rules which must be followed if a private developer or a public authority wants to carry out development in an area covered by the plan. The private property owner decides himself if and when he will develop his property. On the other hand, the plans must be formulated in such a way that they take into account the decision criteria used by a private investor. Otherwise implementation of the plans by private developers is not likely.

PRESENT PLANNING SITUATION

The Urban and Rural Zones Act 1969 formed the first stage of the planning reform. Under this Act Denmark was divided into urban zones, rural zones and zones for summer-cottages. The objectives of this legislation are to ensure planned urban development, to procure a sufficient supply of land suitable for urban development and to avoid harming recreational interests of the population by urban development. It extends powers of nature conservation on which a new law was also introduced in 1969. All areas designated in public and approved plans for urban development or summer-cottages on 1 January 1970 became urban zones and summer-cottage zones. All other areas became rural zones, where only development for agricultural, fishery and forestry purposes could take place without permission. Permission could be granted by the county council for individual cases and by the Ministry of the Environment for the transfer of larger areas from rural to urban (or summer-cottage) zones.

This law has been operated in a restrictive way, especially in the early 1970s, to limit urban sprawl into areas of open country. The Municipal Planning Act 1977 gave the local councils power to approve local plans, which may incorporate proposals to transfer land from rural to urban zones.

It was not until 1973 when the Act on National and Regional Planning was passed that national planning was made mandatory. It is remarkable that the law does not require preparation of a national plan, although national planning must be carried out. This means that the planning process itself is the most important feature of national planning in Denmark. The Minister of the Environment is responsible for national planning. As many other ministers make decisions in matters which are of great importance to the physical planning on the national level, a committee of ministers has been set up to direct national planning. Through the committee any planning issues or major projects with national implications may be evaluated in a broader perspective.

In addition to this co-ordination, the national planning process involves planning directives, an annual report from the Minister of the Environment to Parliament on the development of the national planning work and reports on subjects of special interest in national planning. Also approval and possible alteration of regional plans by the minister may be regarded as part of the national planning process.

National planning directives, which are binding for the county councils when preparing and administering the regional plan, may deal with general goals adopted by Parliament in the context of industrial location, transport, or other policies, or they may be very specific. So far two such directives have been issued. One of them prohibits new zones for summer-cottages in a 1-3 kilometre belt from the coast. The other fixes the location of a future mains transmission network for natural gas.

Among planners and politicians, the annual report from the minister is considered an informal declaration of policy for national planning. One of the most important issues in recent years has been the future urban pattern. To illuminate this planning question the ministry has issued a special report on the urban pattern, and the minister has appointed a commission to examine the role of villages in the future urban structure. The general policy in previous annual reports has been to stop the growth of the biggest towns and to concentrate the population growth in villages big enough to warrant provision of the major public and private services. As proposals for regional plans were being submitted for approval by the Minister of the Environment, the debate in Parliament on the annual report was very intense in the winter of 1980. As a result of the debate, Parliament urged the government to respect the widespread desire to safeguard the future of villages and smaller towns expressed in proposals for regional plans.

The National and Regional Planning Act covers all Denmark with the exception of the metropolitan region of Copenhagen, where one-third of Denmark's population lives. For this region, a special law on regional planning was enacted in 1973. In the metropolitan region a metropolitan council representing all counties and the two biggest municipalities has been set up as a planning authority.

A regional plan for this region was approved in 1979. The plan involves the establishment of a superstructure consisting of a network of 1-km-wide corridors for regional traffic and infrastructure (electricity, oilpipes, gas and water). Along the corridors a belt 1–2-km wide is zoned for activities such as commerce, manufacturing and education. Where these corridors intersect with main traffic lines running radially from the city centre of Copenhagen, it is planned to establish regional centres. Compared with the 'finger' plan of 1947, this established a new structural principle by proposing connections between the 'fingertips'. New urban development in the metropolitan region will take place along the corridors, on the side away from the city of Copenhagen. The regional plan for the metropolitan area was actually prepared in 1967–73, and economic and demographic conditions have changed radically since. It, therefore, remains to be seen whether the goals expressed in the new concept can be achieved even though the economy is in bad shape, and the population in the region is declining instead of increasing.

Outside the metropolitan region the counties had submitted their proposed regional plans by the end of 1979 for approval by the Minister of the Environment. It is a predominant feature of the proposals that a decentralised urban structure is wanted. On the other hand, the counties also want to strengthen the main centres among the smaller municipalities so that a better supply of public and private services may be ensured. In order to safeguard the local services in villages, smaller towns and suburban areas all counties have banned the establishment of hypermarkets, and practically all have restricted the size of new convenience-goods stores. The maximum size of sales area for shops varies from county to county but is in the range 500–3,000 square miles.

The counties show varied attitudes to the question of the extent to which a regional plan should determine the actual location of the building quota allocated to the municipalities. Policies such as these are contained in guidelines for municipal planning, but they must be based on overall strategic considerations and not merely the sum of municipal aspirations.

Regional planning is a continuous process. Every second year the municipal council must review regional development and decide whether there is a demand for changes in the regional plan. As part of the decentralisation process, the county has taken over the operation of several public facilities previously run by the state, such as grammar schools and homes for the mentally handicapped; and some counties have established their own public transport companies, covering the whole region, in order to improve the provision of co-ordinated public transport.

Although the municipal authorities can initiate policies for the regional plan when it is being prepared or renewed, preparation of the municipal plan must not start until the regional plan has been approved by the minister. In the metropolitan region the local planning process started in 1979 and most of the municipalities are at present preparing proposals for municipal plans after conducting public debates on the main planning problems and opportunities. In other parts of the country preparation of municipal plans is not expected to start until 1981-2, but many authorities have been carrying out preparatory surveys, since a plan must be adopted by the municipal council within two years of approval of the regional plan. However, the prepared plan must first be submitted to the county and the minister. The county checks its relationship to the regional plan, and to national planning directives, and if it objects, the local authority may not adopt the plan.

The municipal plan consists of a structure plan containing proposals for the urban structure and development in the urban areas of the municipality, and a recent innovation has been that it also covers the rural districts. Alongside the structure plan guidelines are prepared for housing, institutions, traffic systems, urban renewal and recreation areas for all individual parts of the municipality. These guidelines form a framework for the preparation of local plans. Unambiguous regulations governing this framework were a fundamental condition for decentralisation.

The other type of plan prepared at the municipal level is the vitally important local plan. As part of the decentralisation process, rules for preparation of a local plan have been made more rigorous because local plans transform the structure plan into binding rules for the citizens. A local plan must, therefore, always be worked out whenever it is necessary to ensure the implementation of the municipal structure plan. The municipality has some discretion over whether to prepare local plans, but there are certain cases when it is compulsory, such as larger new developments or allocation of plots, or demolition of buildings. The establishment of infrastructure which will have a major impact on its physical surroundings also requires a local plan. The law gives wide authority to the municipal council regarding the contents and detail of local plans, but the local plan may only be used passively as a restriction in order to control development.

Local plans are registered in the Land Registry for the properties covered by the plan. A local plan is legally binding for the propertyowners in question, unlike the municipal plan which only commits the municipal council and not the propertyowners. Appeals concerning the municipal plan and local plans may be lodged with the minister, but only as regards legal questions and not in matters involving a judgement by the municipal council, since planning decisions are decentralised. Like regional planning, municipal planning is a continuous process. The law stipulates that the municipal council must revise the municipal plan during the first half of each election period of four years.

PLANNING AGENCIES

The planning departments of both the counties and the municipalities have greatly expanded during the last five to ten years as a consequence of the increased demand for planning required by central government and the new planning laws. The planning staff in these departments is still dominated by architects and engineers. These groups constitute 60–65 per cent according to a survey made in 1979. Most of the remainder are people with a non-professional education like technical draughtsmen and clerical and administrative staff. In central government agencies concerned with physical planning, by contrast, about half the planners have an economic or legal education. Physical planning departments are found in various sector ministries, for example, the Ministry of Housing, the Ministry of Traffic and Public Works, the Ministry of Energy and the Ministry of Education, as well as in the Ministry of the Environment.

Up to the early 1970s it was common for major tasks like the preparation of a structure plan to be executed by consultants. Most of the consultants were architects covering all fields of physical planning, but there were also consulting firms specialising in planning for traffic, commercial centres and landscape. In connection with the expected increase in urban renewal activity both the association of non-profit-making building societies and the national association of propertyowners have established special urban renewal companies.

With improved staffing of planning departments in recent years in the counties and municipalities, a concentrated planning effort to implement the new planning laws and a society which is expected to change more slowly in the coming years, the role of the consultant will be changed. Instead of doing most of the planning work, consultants will be used for specialised planning subjects and to contribute with their broader experience of different planning situations.

CONCLUSIONS

With planning reform, far better planning instruments have been created. The objective of decentralising planning decisions, and increasing citizen participation in planning, have been met with enthusiasm in some quarters and scepticism in others. So far the experience with the new legislation is limited, so it is not possible to evaluate it fully.

As far as citizen participation is concerned regional planning offers some evidence. Analyses in a couple of counties show that about 5 per cent of the adult population have participated in one or more meetings or other activities, and 30-35 per cent had a knowledge of one or more of the subjects dealt with. Considering the very limited number of people participating in the public debate on other subjects the results are acceptable. In general, the participation rate is greater when it is a matter of local affairs, and an increase in citizen participation must be expected as the preparation of local plans assumes greater prominence when the municipal plans have been adopted.

The decentralisation of planning decisions is taking place within a framework of general regulations and rules set up by central government agencies. It is not yet possible to judge whether the guidelines, as they are administered, are giving enough latitude for the municipalities and the counties to form their own policies. However, the Ministry of the Environment has stated that revised advice on the interpretation of the laws will be forthcoming when more experience with the new planning legislation has been gained.

Chapter 11

Greece

A.-Ph. Lagopoulos

This chapter is a brief but systematic account of the system of physical planning in Greece, as it existed in August 1979, that is, as it has been shaped under the right-wing government of the country. The text was then completed with some important later developments under the right-wing regime, and significantly updated in November 1982, approximately a year after the election of the PASOK (Panhellenic Socialist Movement) government. Thus, the use of the present tense refers in general to the system of 1979, while the later developments are accompanied by their respective date. It is important to note, however, that the evolution of the Greek planning system from 1979 to the present has been such as in no way to alter the general image of it given in this chapter. The basic organisation of the material is as follows. Initially, the periods and changes of the Greek social formation during modern times are outlined, and related to state intervention and regional and urban planning problems. Then, the structure and function of the physical planning system is considered, as well as the role of the relevant agencies, local authorities and private practice. Legislation is presented next, noting that the foundations for its modernisation were laid during the 1970s. Finally, the theories, methods and policy of those involved in planning are discussed.

GREEK SOCIAL FORMATION AND PLANNING PROBLEMS

The history of modern Greece, which begins with the liberation from Turkish domination (1827), can be structurally divided into three main periods. The first period (1827 to 1907–13) is characterised by the spread of mercantile capitalism in relation to feudalism. The state was initially largely controlled by the foreign royal court, but underwent some modernisation towards the end of the period. The second period (1907–13 to 1950–3) represents a period of liberal capitalism.

Industrialisation in essence began, and developed slowly, accompanied by a rise in the proportion of workers in the population (around 5 per cent in 1928). At the beginning of the period this group began to form an identifiable social class. Political instability was also a feature of this period, culminating in the Civil War.

The third period (1950-3 to the present) corresponds to 'monopoly' capitalism. The bourgeois class (the large capitalists) is very small, and while itself dependent on international centres, controls Greece economically and politically, through right-wing governments or even military dictatorship (1967-74). There is also a middle class, quantitatively dominant, and an important working class. The proportion of manual occupations in the total active population was 30 per cent in 1971. The development of capitalism has been accompanied by a much more balanced composition of the economically active city population, compared with the prewar one, and by a rapid rate of industrialisation and modernisation after 1960. British political influence in Greece has been followed after the Second World War by direct American political and economic influence, and today the country has become a full member of the EEC, after an associate membership of about twenty years.

During the nineteenth century the areas of highest population density were distinct from the most urbanised areas; but since the early twentieth century they have become coincident, demonstrating the increasing dependence of economic development on urban poles. The urban hierarchy was relatively simple and limited in the nineteenth century, but during the twentieth the urban hierarchy became steadily more complex and elaborate in association with industrial and economic development. One consequence of this process is an increased opposition between town and country. The urban hierarchy is dominated by the two major cities, the Athens/Piraeus conurbation, and Thessaloniki.

During 1951-71 the population of Greater Athens increased from 1.4 to 2.5 million approximately, and that of Greater Thessaloniki from 300,000 to 550,000 approximately. Third comes Patras, with around 120,000 population in 1971.

Over 1920-71 the formal rural population, in settlements with less than 2,000 inhabitants, decreased from 62 to 35 per cent; the semi-urban population in settlements with 2,000-10,000 inhabitants decreased from 15 to 12 per cent; and the urban population in settlements with more than 10,000 inhabitants increased from 23 to 53 per cent. The total population in 1971 was approximately 8.8 million.

Regional and urban problems arise from the aspatial and spatial structure of the social formation. The main regional problem is a marked regional inequality, with very great concentration of productive forces, decision-making, cultural activities and population in Athens and, secondarily, Thessaloniki. This contrasts with backward agriculture, widespread underemployment and low income and quality of life in the greater part of the periphery. The most problematic parts of the periphery (approximately 50 per cent of the country) are mountainous, insular and frontier regions. Another main problem is the conflict and lack of organisation among land uses. A serious environmental problem, in addition, is created by the large-scale degradation of the physical environment in the whole country.

The dominant urban problem since 1922 has been housing. In the postwar period this was due mainly to the destruction during the war and to regional inequality, while today it is due mainly to the latter. A part of the problem, illegal construction (not to be equated with slums) outside the area of the official city plan of Athens and Thessaloniki, and other large cities, appeared with the arrival of the Greek refugees from Asia Minor starting in 1922. It became a large-scale phenomenon after the war and today dominates the housing problem. Recently total illegal constructions in Athens were probably of the order of 150,000.

The housing problem is, in part, a problem of housing quality which is most acute in Athens and Thessaloniki. It is mainly the result of lack of organisation of the urban environment especially in illegal construction areas; uncontrolled extension of housing areas; the very limited public open and green areas; all kinds of environmental pollution; the disadvantages of the types of housing development used; and the high total floor area/plot area ratios and densities in central areas resulting from the intensity of land exploitation. About ten years ago mixed densities of up to 600 persons per hectare were found in Athens, and net densities of up to 2,000 persons per hectare existed in Thessaloniki as recently as twenty years ago. The problem of housing quality in the declining settlements of the periphery consists in the very poor quality of the dwellings and infrastructure, and in the gradual abandonment and deterioration of dwellings, many of which are of architectural value. Finally, summer-cottage construction is often illegal and unplanned, frequently to the detriment of the natural environment.

The development of city centres seems to be considerably retarded in comparison with the cities of Western Europe and the USA. Only Athens and Thessaloniki have a significant central business district. The CBD is the only area of significant concentration of economic and administrative uses. Both this and other concentrations elsewhere in the city result mainly from private initiative, without deliberate planning. Development of the central city has been accompanied by the destruction of the historic character of its older parts and by poor functional organisation. Considerable traffic problems also exist in the city centres.

Industry is scattered in an unplanned manner in and around the cities, and planned industrial parks are very limited. Generally a major problem is the conflict and the lack of organisation among land uses.

If we consider the urban infrastructure, the road network in the cities has been modernised only to a very small extent and is saturated in the central areas. Service by public transport is poor, while the number and use of private cars is increasing steeply. There is an almost total lack of proper drainage (one exception is about 30 per cent of the Athens area), and one result of which is the periodic flooding of roads and basements. In addition, the dumping of sewage and industrial wastes with no previous processing creates serious pollution of the natural environment. The remaining technical services are in a more satisfactory condition, though they are frequently inadequate for modern needs.

State intervention in regional problems became significant in the middle of the first period, referred to earlier, and for urban areas, neoclassical designs were drawn up then, but played a primarily ideological role. During the second period spatial planning became more rationalised, primarily however on the theoretical level. The policy of fragmentary city plans (see below) began, and an attempt was made to rehouse the refugees from Asia Minor. At the end of this period the state undertook small-scale public works, while housing was undertaken by private capital. This continued during the third period, when private capital undertook significant real-estate investment. Housing production by private capital is oriented towards single buildings and not towards housing complexes. This period is characterised by increased rationalisation, generalisation and specialisation of physical planning on the administrative level, increased attempts at state intervention expressed through the administrative machinery and legislation, and the production starting about 1960 of regional and master-plans which were not implemented.

Even today, state intervention is limited and fragmentry, and often produces results contrary to those suggested in official rhetoric. State intervention is likely to increase in the future, within the framework of market mechanisms, and will become more comprehensive in order to balance the conflicting interests of capital, and provide the infrastructure necessary for its reproduction. Emerging popular pressure, and demand for social infrastructure, will also influence the development of state intervention.

SYSTEM OF GOVERNMENT

Greece is a presidential republic. Executive power is vested in the government and administration. Each sector of administration is assigned to a ministry, which is relatively independent, and is responsible for planning in its sector and the execution of policy. Alongside these there is one ministry, the Ministry of Co-ordination, whose object is intersector co-ordination of programmes and allocation of the funds of the public investment programmes.

Greece is divided into a four-tier geographic hierarchy. The largest geographical division is the province, which is not in general an administrative unit. Next comes the department (*nomos*), of which there were fifty-one in 1977, administered by a prefecture, in which most of the ministries are represented. The third level is the county (*epoucchia*), of which there are 147. Finally, the municipalities and the communes, of which there are 264 and 5,759 respectively, are administered as local authorities supervised by the prefectures and the Ministry of the Interior.

During modern times the prevailing tendency has been towards centralisation of administration. This is true also of recent attempts at modernisation, in spite of the government's stated intention in January 1980 to 'complete' the process of decentralisation. The relationship of local government to central government follows the same principle of centralisation. The basic law referring to local government today (LD 2888, 1954), as well as the constitution, states that among its responsibilities is the administration of local affairs, but this term is interpreted in a limited sense. The state controls many of the activities of local government, and may dismiss its elected members. Thus, local government functions as a direct extension of the state. It has no economic self-sufficiency and is not even in a position to fulfil its economic responsibilities. In January 1980 the government did decide to grant greater independence to local authorities, but this decision is not implemented. A great emphasis on decentralisation and local authorities is given by the new PASOK government, but the practical results are not yet evident.

The legislative organ is composed of Parliament and the head of state, the president of the republic. Formal laws are passed by vote of Parliament and ratified and issued by the president. They are based on the constitution, which constitutes the fundamental law of the country. The president also issues presidential decrees (PDs), which are countersigned by the appropriate ministers. Among these are the legislative decrees (LDs), which are equivalent to formal laws and issued under exceptional conditions; the regulative decrees, issued in implementation of formal laws; and the organisational decrees, referring to the organisation and function of public agencies and corporations.

RESPONSIBILITY FOR PLANNING

The main organs of physical planning are certain ministries and public agencies. First, the Ministry of Co-ordination (from 1982 the Ministry of the National Economy) will be considered. It is the ministry responsible for general policy in the areas of regional physical planning and environmental protection, and for regional planning studies. Until a few years ago it also prepared master-plans. Important responsibilities are co-ordination of the preparation and realisation of national programmes of economic and social development; economic foreign policy; and general national policy in view of the integration into the EEC. Since 1964 it has supervised the Centre of Planning and Economic Research (KEPE), and is responsible for the national statistical service. Of the total professional staff, about 70 per cent are economically oriented, and no posts are specifically created for physical planners.

The Ministry of Co-ordination includes the Public Investments Agency, the Agency for Regional Policy and Development and nine Agencies of Regional Development which play a secondary role. The Agency for Regional Policy and Development is primarily responsible for the preparation of the programmes of regional development; it also prepares the programmes to be proposed for financing by the European Regional Development Fund. In 1979 the Agency had been occupied for two years with programmes of regional development mainly referring to selected settlements with limited population, the effectiveness of which will be quite limited. In autumn 1982 the agency was gathering reports from its peripheral agencies, parallel with the invitation by the ministry to various persons and agencies to undertake development studies, as input for the new five-year plan. The ministry also includes the Agency for National/Regional Planning and the Environment, responsible for the preparation of national and regional planning policy and programmes and their implementation. The main goal of this agency in 1979 was the rushed preparation of a national plan. A similar programme was also being prepared independently by KEPE, tailored to the requirements of the European Fund for the financing of projects in underdeveloped regions. Environmental-protection policies, and the preparation and implementation of the planning of water and land resources, are also responsibilities of this agency. The agency employs about 120 professional staff concerned with regional development.

Similar responsibilities for environmental protection in residential areas and of the major ecological environments are held by the Ministry of Social Services. In the framework of governmental collaboration with the United Nations this ministry is preparing a programme for environmental pollution control in Athens.

In association with the Ministry of Co-ordination expert committees are appointed, responsible for reports, advice, or implementation in matters of physical planning. Typically they include university professors or the Technical Chamber of Greece. The Minister of Co-ordination is president of the National Council of National/Regional Planning and the Environment, which is responsible for decision-making on general physical planning and environmental subjects and for supervising the implementation of such programmes.

A decision was taken a few years ago to found a new ministry, responsible for the preparation of regional planning studies and environmental programmes, and with decision-making power over urban programmes and housing projects. After a delay due to conflicts of interest at ministerial level, the ministry came into being in the beginning of 1980, with the title Ministry of Regional Planning, Settlement and the Environment, and is based on the General Settlement Division of the Ministry of Public Works (see below). In 1982 this new ministry sent its employees to the country departments to gather empirical and piecemeal information on regional issues and problems. This information is intended to feed the new five-year plan, and to contribute to Operation Urban Reconstruction, 1982-4. This operation, which is intended as the first phase of a wider urban reconstruction, concerns general urban plans and urban studies (see below), as well as ordnance and topographic surveys of approximately 400 cities, towns, villages, and municipalities and communes of Athens and Thessaloniki.

The operation, which was announced a week before the municipal elections, demands interdisciplinary study groups and aims at decentralisation in the elaboration of the studies. The latter are to follow a series of given prescriptions. The whole operation was hastily mounted and provoked a feverish reorganisation of the planning offices. The practical exclusion of the universities and the indiscriminate inclusion of unqualified persons creates serious doubts about the quality of the future studies. Worse, the studies, with a cost of the order of £10 million, will be planning in a legal void or with a provisional planning law (see below), and without any real mechanism, or to our knowledge any financing, for their implementation.

Since 1957 five-year national-development programmes have been prepared by KEPE. The most recent, for 1978-82, was adopted in 1979. This plan is typical of Greek plans, except for a greater focus on physical planning than was the case previously. It is characterised by a lack of theory or coherent methodology, and is based on fragmentary empirical data. It is largely utopian. These five-year plans have been of very limited effect, for these reasons, and also because of a lack of mechanisms and economic resources for implementation.

This problem also affects the regional planning studies, none of which have been given legal status. Such studies are used in an informal and *ad hoc* way. They are prepared by the Ministry of Co-ordination, together with other public agencies and private firms, since approximately 1960, and cover an important part of Greece.

Urban planning matters were until 1980 the responsibility of the Ministry of Public Works; it is still responsible for the road network and public works, map-making (for which the Army Geographic Agency is also responsible), and the compilation of the national land register. Its professional staff numbers around 2,000, of whom about 85 per cent are qualified in technical subjects. Posts are not specifically reserved for physical planners.

The General Settlement Division, which today belongs to the Ministry of Regional Planning, Settlement and the Environment, is responsible for urban master-plans and housing policy. Much of the work of preparing plans is, however, contracted out to private firms of consultants. None of these plans, which cover a considerable number of settlements, has been implemented or legally adopted. This is partly because of the lack of procedures to secure implementation, and lack of economic resources. For example, compulsory purchase creates serious problems as a result of high land values. Like the regional studies, master-plans are used in an informal and *ad hoc* way. In 1979 a series of studies reviewing the master-plans for the cities identified as regional growth poles (see below) were commissioned from private firms.

Two examples demonstrate the adventures of the master-plans in Greece. The agency responsible for the Athens master-plan produced a proposal on which opinions differed widely. For this reason, the ministry requested five different consultants to prepare planning proposals within forty days. This process by-passed any real participation by the city's political representatives. The end-products were patched together by the ministry into a set of proposals. In 1979 contracts were assigned to nine private firms for the study of nine zones into which the department of Attica has been divided, but no provision was made for real co-ordination of these studies.

In the case of Thessaloniki, a master-plan was prepared in 1966. After two additional proposals, the city, the university, the Technical Chamber and a private firm were asked to make their own proposals, on condition that they do not co-operate. Nevertheless, they succeeded in working together, and argued the case for a special planning agency for the city.

The same division is also responsible for preparing and approving city plans. City plans are the only type to be implemented so far. However, this does not imply that proper planning principles are followed. Very often, the implementation of these plans legalises development initiated by speculative developers. The extension of cities is tied to a set of fragmentary city plans, the number of which for Athens alone, exceeds 2,000.

The same division is also responsible for the planning of zones of active urban planning (ZEP, see below), of which none has as yet been fully implemented; the planning of housing and the preparation of housing projects, which frequently concern disaster relief and generally small settlements or limited areas of settlements; the evaluation and protection of traditional settlements; and the drafting of measures for the protection of the environment from pollution. Comprehensive urban-improvement studies have been prepared for what is left of the old inner cities of Athens and Thessaloniki. Proposals for the latter are encountering strong reactions for the inhabitants. Expert advisory committees on urban planning are appointed in association with the responsibilities of the ministry.

Before the creation of the Ministry of Regional Planning, Settlement and the Environment, the Ministry of Public Works supervised the Public Corporation for Housing and Urban Planning (DEPOS), which is supposed to intervene in housing and urban conditions, generally through the planning of urban complexes, but has not as yet been able to fulfil this purpose.

Protection of traditional settlements is also a responsibility of the Ministry of Culture and Sciences, together with protection of archaeological sites. The conservation of modern monuments is quite limited. In 1972 the Ministry of the Interior listed about 2,200 traditional settlements according to three categories of protection, but in 1979 the lack of protection had resulted in about 400 settlements losing their traditional character.

The Ministry of the Interior is also responsible in various ways for the administration of the planning system. Each departmental prefecture includes a Planning Agency, which is formally responsible for co-ordinating the departmental agencies responsible for implementation of regional plans. The ministry also operates the departmental Technical Services of Municipalities and Communes, which on paper undertake plan preparation on behalf of local government. The prefect generally has formal responsibility for approving city plans for settlements of up to 5,000 inhabitants. Local authorities themselves perform a minimal role, more commonly advisory in a general way, in planning, due to their lack of power and resources.

Other government departments also have physical planning responsibilities. The Ministry of Agriculture is concerned with agricultural planning, forests, forest roads, landscape and planning for settlements of up to 4,000 inhabitants, in the context of the resettlement of refugees and landless peasants. There are about 3,000 such settlements. The ministry is responsible for ten areas, designated national forests, including one on Olympus, and about twenty designated aesthetic forests.

The Ministry of Industry and Energy is responsible for industrial-location policy, and the planning of industrial estates. Their organisation and exploitation is undertaken by the state-owned Hellenic Industrial Development Bank (ETVA), which in 1979 established a separate agency for this purpose. Twenty-nine industrial estates were planned, of which five were operating, three were beginning to function and six were being developed by 1979.

The National Tourist Organisation of Greece (EOT) is responsible for tourism policy, consultation and preparation of tourism studies. It also designates tourist sites and zones.

In the field of housing the Ministries of Public Works and Social Services, the Autonomous Organisation for Workers' Housing supervised by the Ministry of Labour, the Autonomous Building Organisation for Officers and the National Mortgage Bank of Greece SA, provide housing loans and housing projects for certain social groups. The housing activity of the public agencies makes up a very small percentage (about 5 per cent for the 1960–72 period) of the total housing production, while the credit policy of the government results in housing-financing accounting for about 20 per cent of gross fixed-capital investment in housing.

The physical planning system in Greece exhibits certain major defects, including lack of effectiveness, excessive centralisation (which even the government admits) and authoritarianism. There is also a lack of real participation by regional administration and local government. Overlapping and fragmented responsibilities are recognised by the government and are frequently the subject of criticism, but lack of real co-ordination is a more serious defect.

Centralisation and authoritarianism are demonstrated by considering the form of participation of local government, non-government agencies and individual citizens. The model for such participation is still a law of 1923 on city plans (see below), which grants local authorities an advisory role and gives citizens the right to be informed and lodge complaints. This amounts to a right to make a more or less ineffectual protest. Thus, there is a pseudoparticipation in planning, although preliminary contact with the citizens, *de facto* imposition of urban organisation by capital or low-income groups, and public pressure and action are common.

URBAN AND REGIONAL PLANNING LEGISLATION

Until about ten years ago urban planning legislation was very limited, in spite of its complexity. It had a regulative and static character, was concerned with narrowly legal and technical aspects and was indistinct from the General Building Code. Only in recent years have laws providing a wider view of planning been enacted, but urban planning legislation is still largely outdated, while regional planning legislation is largely non-existent.

Under the constitution planning is controlled by the state. Private property is protected, but social implications of property are also recognised, since the constitution states that rights to property cannot be exercised to the detriment of the common interest. The constitution also introduces the concept of Planned Development Area.

The basic law on planning was until recently the legislative decree of 17 July 1923, 'concerning the plans of cities, towns and communes of the state and their construction', which was surpassed on paper by legislation only after 1970. This law was advanced for its time, and provided for every settlement to be regulated and to develop in accordance with a 'city plan', which defines public open spaces and spaces where construction is permitted, and a very limited number of land uses. These plans are subject to constraints governing plot ratios and dimensions, site coverage, height of building, type of housing development, and so on. Expansion of the built-up area, with special functional constraints, in a zone of 500 metres around the city was in principle permitted. A later presidential decree in 1928, 'concerning the definition of conditions and constraints on the construction of buildings within and outside the zone of cities', defined those constraints.

The law of 1923 embodies an outdated conception of urban intervention. The city plan is a narrow and static concept and amounts to no more than a regulation of individual exploitation of the land. The basis for a more modern approach to planning began to be created in the 1970s. The concept of the city plan was widened and given a dynamic and comprehensive character; emphasis was given to large-scale planned construction; the master-plan and a dynamic regional plan were defined; and the state affirmed its intention to intervene actively in urban and regional matters.

The first law of this modern series was legislative decree 1003 of 1971, 'concerning active urban planning'. It was based on French, and to some extent, German models. It is oriented towards comprehensive urban planning, incorporating physical, economic, social and aesthetic factors, for zones designated for development or improvement (ZEP). The planning study for a ZEP, which is prepared by the construction agency, must include all these factors, and take into consideration the area around the ZEP. The construction agency can be mixed, that is, its capital can originate from private as well as public sources. A ZEP is proclaimed on the initiative of the Minister of Public Works. The approval of the study follows the law of 1923. Thus, no real provision is made for citizen participation.

A basic defect of law 1003 is that in practice it potentially allows significant profiteering by big capital. Another serious defect is that no provision is made for a plan covering a wider area to provide a context for the ZEP. This point was covered by a law in 1972, legislative decree 1262, 'concerning master plans for urban areas', which defined the concepts of 'regional plan', 'Master plan', 'urban plan' and 'land use'. The regional plan determines the spatial organisation of a series of social (in the wide sense) data in the context of development programmes. The master plan refers to the regulation and control of land use and urban networks in the context of wider development programmes, or in their absence, of the national/regional governmental policy. It includes a survey of existing conditions, on which the justification of the proposal should be based, forecasts, goals and targets of development, and proposals for implementation. Initiative for approval rests with the Ministers of Public Works, Co-ordination and the Interior. Local government acts in an advisory capacity, and a procedure is provided for lodging complaints, but not by citizens.

The law which was intended to replace the 1923 law is 947 of 1979, 'concerning planned development areas', which supersedes and extends legislative decree 1003, but has been suspended by PASOK. It analyses the concept of a planned development area (OP), which is an area where development can take place, based on the general principles of urban planning and any regional plan that may exist, and in a form consistent with social or economic goals. The OP is not explicitly related to master plans, which is retrograde in comparison with the 1972 law, but is from a certain point of view realistic.

An OP requires a study including a survey of existing conditions, forecasts of population and needs for constructed space, a simple estimate of the environmental consequences of the proposal, the proposed conditions and manner of development or improvement, the stages of implementation, and the costs and benefits. General land uses, and exceptionally, specific uses, plot ratios, and so on, are defined. Before adoption, appropriate government departments may intervene, and the plan is placed on public display. Evaluation of objections is the responsibility of the Minister of Public Works, on whose initiative the OP is proclaimed. After proclamation of an OP, compulsory purchase within it is possible.

Three manners of development or improvement of an OP are defined. The first is active urban planning, defined as in legislative decree 1003; the second is urban land consolidation, according to which the propertyowners collectively contribute their property in return for a real-estate concession; and the third is through the establishment of normative building regulations. Active urban planning can be undertaken, as in law 1003, by partly publicly owned agencies. In this case, the public share must exceed 34 per cent.

Propertyowners, whose participation in the planning process is still rudimentary, are required to concede a proportion of their plots for the organisation of the public open spaces, and to make cash contributions for the construction of the basic infrastructure. These measures reinforce the effectiveness of the state, but also that of big capital. They are based on sound principles, but in their specific contents they discriminate against small property, and in practice they would create such social problems that the new law is now being rewritten. This law represents an attempt at modernisation, in preparation for EEC entry and in order to provide for mass production of housing. It potentially favours the entry of foreign capital and technology, and has been widely criticised by citizens, local government and the technical professionals in Greece.

In the summer of 1982 a proposal for a transitional urban planning law was prepared by the PASOK government to replace law 947. The main concerns of this proposal are urban extension or incorporation, and illegal housing. The main theoretical concept for urban organisation is the idea, of questionable relevance, of the neighbourhood.

There are two levels of urban proposals and plans: the (general) urban plan and the urban study; the former is similar to the master plan of law 1262, but impoverished in that it is isolated from the regional frame, and the latter to the city plan of the General Building Code of 1973, while giving a greater emphasis to implementation. Active urban planning and urban land consolidation of law 947 are retained, while the concepts of zones of special reinforcement and zones of special incentives are added. For the first two, the public share in the case of partly publicly owned agencies must exceed 50 per cent. The law proposal provides for contributions in land and cash, but contrary to law 947, the first is proportional to plot size.

Concerning illegal housing, houses constructed after 10 December 1981 are demolished. Those constructed before that date can be legalised with a cash payment.

Citizens' participation in the preparation of the plan must be encouraged, according to this proposal, and the neighbourhood urban committees are elected by the inhabitants. But with all the explicit encouragement and enlargement of participation, the model of the law of 1923 has not been superseded.

The Greek Architects' Association considers that the above law is a positive contribution, but must be considered as a provisional law of restricted application, to be extended in order to apply to the whole of the country. We think that this view very accurately reflects the provisional character of the proposal.

In the field of national and regional planning, and environmental protection, the basic law is 360 of 1976, which provides for state intervention at the larger spatial scales. It contains definitions of the concepts of 'national and regional plan' and 'programme', and of environmental concepts; it regulates preparation and approval procedures; and establishes relevant agencies. City plans and master plans must be consistent with the national and regional programmes prepared under this law. Regional economic development legislation is closely related to national and regional planning. Incentives for industry to develop in peripheral areas were first adopted in 1949, and since then there has been a series of relevant laws (for instance, in 1971 and 1978). The policy, however, has been a failure, since industrial development remains concentrated around the two major cities, while a very large part of the country is undeveloped to a striking degree. The first incentives for tourist development in the periphery were adopted in 1972, but this also remains concentrated in few areas.

In 1982 law 1262, 'for the granting of incentives for the economic and regional development of the country', was issued by the PASOK government; it aims at the encouragement of productive investments in the primary and secondary sectors, modern technology, transportation of persons and products, and tourism. The regions of the country are divided into four categories, from the most to the least developed, and a variety of incentives is given, the importance of which increases in proportion to the decreasing degree of development.

EDUCATION, THEORIES AND POLICY IN PLANNING

Greek physical planning has a limited history of theoretical development, and almost no history of practical experience. Much of its early history springs from the personality of the architect C. A. Doxiadis, who by 1940 was already directing an agency for urban and regional planning studies for the Ministry of Public Works. The first chair of urban planning was founded in the School of Architecture of the National Technical University in Athens, a chair which until the early 1960s had a very limited and theoretical orientation. This chair and the private activities of Doxiadis based on grandiose ideological theories, and a few public agencies, constituted the core of physical planning during the 1950s. During the next decade were added a second chair of urban planning at the School of Technology of the University of Thessaloniki, KEPE and ETVA, and a number of private offices. More appeared during the 1970s, and now there are five university chairs of physical planning.

Since there are very few university chairs in human geography, and since the study of spatial phenomena is underdeveloped in the academic fields of economics and sociology (the latter is practically non-existent in the country's universities), Greek physical planners come almost exclusively from an architectural background either through hypothetical professional experience or postgraduate studies abroad, mainly in France, Britain and Germany. The influence of this background is especially evident in earlier planning studies.

The theories underlying the first generation of studies during the early 1960s show the same fundamental weaknesses which also mark urban and regional planning internationally during this period: empiricism, projection of the architectural approach into larger spatial scales, exaggerated emphasis on physical space and its possibilities (a kind of spatial determinism), a tendency to ignore aspatial data, a functionalist approach to land-use organisation and the application of 'tree-structure' logic to spatial organisation. These theories tended to be apolitical and technocratic, and the studies were further weakened by the absence of any real legal framework for implementation, and a corresponding absence of realistic orientation.

The second generation of studies approximated to the period of the military dictatorship, while a third generation is in the process of emerging. There has been a reduction of empiricism, and a greater concentration on the aspatial factors of the social formation, coupled with a strong politicisation, especially after the dictatorship, of a left-wing orientation. Wider participation in plan-making has become a central issue. Another main characteristic of today's views is the emphasis on housing quality and the environment. In contrast to the earlier cartographic, end-state approach, planning today is considered to be a continuous, spiral process. And there is a strong preoccupation with the problems of implementation.

The overall objectives of the most recent five-year national plan, and a decision of the National Council for National/Regional Planning and the Environment, both in 1979, give the only official view of planning policy. This includes restriction of development and population growth in Athens, Thessaloniki and the main tourist centres; the development of nine regional growth poles and of probably more than fifty significant settlements, and of a number of rural settlements around which the rural-settlement network is to be organised; urban improvement for the two main cities; and protection of the natural environment and cultural heritage. This overall plan seeks for the first time to integrate a whole range of sectoral policies into one programme which is generally and theoretically desirable. One reason for attempting this is to take advantage of EEC aid to peripheral regions of the Community.

However, this attempt presents a series of weak points. It is *ad hoc*, and less global than it may appear. The policy of development poles, based on a loose interpretation of Perroux's theory, has been a failure internationally. Furthermore, the proposed model of development attempts a utopian break with the present situation, since it demands reduction of the development of Athens and Thessaloniki, and the transformation into regional poles of cities which generally do not fulfil the necessary conditions for this. The regional problem in Greece has its source in the development of capitalism, and the representatives of capitalism were not in a position to confront it either in theory or in practice.

CONCLUSIONS

The planning system in Greece is oriented towards the resolution of the regional and urban problems created by the particular form of dependent capitalist development to which Greece has been subject. During the last few years an attempt has been made at greater modernisation of the physical planning system, and urban and regional planning legislation, mainly in view of the country's integration into the EEC. The ministry responsible for the fields of national and regional planning and the environment is the Ministry of Co-ordination, while the ministries responsible for matters of urban planning, public works and the road network are the Ministry of Regional Planning, Settlement and the Environment, and the Ministry of Public Works. Apart from these two, several other ministries have some responsibility for planning: the Ministries of the Interior, Agriculture, Industry and Energy, Social Services, and Culture and Sciences, as well as the public agencies and the local authorities.

Main weaknesses of the planning system are ineffectiveness, excessive centralisation, lack of real co-ordination, and lack of real participation by non-government agencies or individual citizens. The mechanisms and economic resources for implementation are lacking, and no regional planning study or master plan has been statutorily adopted.

Until ten years ago urban planning legislation was exceptionally limited, and regional planning legislation was non-existent. The framework for more modern urban and regional planning legislation began to emerge during the 1970s, but in spite of this, Greek physical planning legislation is insufficient and in need of modernisation.

With the new legislation, planning projects are given a dynamic and comprehensive character and the state intervenes actively in the regulation of space. Nevertheless, this legislation has not yet been put into practice, and urban development is still tied to basically geometric city plans.

During the early 1960s the theoretical base of planning tended to be apolitical, technocratic and empiricist, a fact reinforced by the architectural background of most practitioners. Politicisation and theorising has come since, with a turn towards the aspatial factors in the production of physical space. On the other hand, the 1979 official model for planning, as expressed in the national planning policy, is utopian, and based on the vague concept and internationally unsuccessful policy of development poles.

Chapter 12

The European Communities

R. H. Williams

The movement towards European co-operation and ultimately integration in the early postwar period was to a large extent motivated by the desire to prevent the possibility of the outbreak of another European war. Much of the initial drive came from Jean Monnet, postwar head of the French National Planning Authority, and Robert Schumann, French Foreign Minister, who sought to replace the historic rivalry between France and Germany by a system of mutual economic and political co-operation, embracing other nations as well, which would make a future war 'not only unthinkable but materially impossible'.[1] Although the initial impetus strongly reflected the desire to achieve political objectives, economic aims have always been prominent in the minds of those promoting European institutions, particularly after some of the initiatives towards political and military integration in the early 1950s lost their impetus. The desire to create a larger free-trade area than any individual European nation could achieve on its own was an important element in the European movement. The customs union between Belgium, the Netherlands and Luxembourg had been implemented in 1948, followed by the establishment of the European Coal and Steel Community (ECSC) in 1951. The latter included all the six nations who later became founder-members of the European Economic Community, namely, France, Italy, the Federal Republic of Germany, the Netherlands, Belgium and Luxembourg.

In June 1955 the Foreign Ministers of these six countries met at Messina, in Italy, to discuss the possibilities of more general economic integration. Following this meeting a committee chaired by Paul-Henri Spaak, Foreign Minister of Belgium prepared a report which formed the basis for the Treaty of Rome, establishing the European Economic Community (EEC). This treaty, together with that establishing Euratom, was signed in Rome on 25 March 1957 by the six member-states of the European Coal and Steel Community. Thus, the EEC and Euratom came into being on 1 January 1958, creating the three

The European Communities 145

European Communities that we have today. These three communities were enlarged, after a protracted period of negotiation, on 1 January 1973 by the addition of the UK, the Republic of Ireland and Denmark. A further enlargement took place on 1 January 1981 with the addition of Greece, and another is in prospect if the applications from Spain and Portugal to join are successful. Full accounts of the history, objectives and policies of the European communities may be found among the references listed in the EEC section of the Bibliography.

Consideration of the original ambitions of the European Communities might suggest that they would not have a great deal to do with town and country planning. However, as this chapter seeks to indicate, there are a wide variety of ways in which matters of concern to town and country planners are connected with aspects of EEC policy. This is clearly evident if one considers the range of issues likely to be the responsibility of a planning authority, including environmental protection, investment in urban and transport infrastructure, and intervention in the economy to reduce regional inequalities and alleviate the problems of areas of high unemployment. Indeed, as Sir Meredith Whittaker said in 1979, 'There is no major sector of local government that is not affected by what is decided in Brussels'.[2]

In this chapter the main Community institutions of relevance to various aspects of planning, and the means by which they exercise their authority, are outlined. This is followed by a review of the major areas of EEC policy-making and action that are of concern here. As indicated above, there are three European communities, that is, the European Economic Community (EEC) or Common Market, the European Coal and Steel Community (ECSC) and Euratom. The three communities are administered together in Brussels by the Commission of the European Communities. For convenience, the short titles of the Commission and the EEC are used here. Euratom is not of direct concern, and the ECSC is referred to specifically where necessary. Otherwise, the EEC is the Community referred to.

COMMUNITY INSTITUTIONS

The major institution to be considered is the Commission of the European Communities. This is not, however, the ultimate decision-making institution, since this power rests with the Council of Ministers. Both the Commission and the Council are advised by the European Parliament and, for EEC and Euratom purposes, the Economic and Social Committee.

The Commission consists of fourteen members, appointed by agreement with the national governments for a four-year term of office. Germany, France, Italy and the UK appoint two each, and the other countries provide one each. Members of the Commission are assigned responsibilities by the president of the Commission, and during their term of office they are expected to act in the interests of the Communities as a whole, and not on behalf of their own country. The Commission is responsible for ensuring adherence to, and acting upon, the provisions of the Treaties of Paris and Rome which set up the Communities; it is the initiator of Community policy, and is responsible for enforcing Community laws once adopted. It is the executive arm of the Community; and it represents the Community interest to the Council of Ministers. Each member of the Commission has one or more directors-general reporting to him, in accordance with the responsibilities he has been assigned. The directors-general are the chief officials of the twenty directorates into which the administration is divided. The directorates-general (DG) that are of relevance to aspects of planning include: DG III, Internal Market and Industrial Affairs, whose responsibilities include recognition of academic and professional qualifications, and freedom to practise throughout the Community, and industrial structure, especially in the context of the steel industry; DG V, Employment and Social Affairs, responsible for the Social Fund; DG VII, Transport, responsible for developing a common transport policy; DG XI, Environment, Consumer Protection and Nuclear Safety; and DG XVI, Regional Policy, responsible for the European Regional Development Fund (ERDF). DG XI is a directorate-general of great importance for planning, because it has responsibility for the Community's Environment Programme, which now places considerable emphasis on land-use planning, and includes a number of planning measures, as well as a more general responsibility on behalf of the Commission, for urban problems, urban policies and town planning. This directorate-general was created in 1981, and its responsibilities were previously carried out by the Environment and Consumer Protection Service. It now has equivalent status to that of the other nineteen directorates-general.

The total staff of the Commission of the European Communities is not large when compared with many national government departments: 8,121 in 1978, of which 2,116 are in administrative and executive grades.[3] Most of these are located in Brussels, but about 1,700 officials are based in Luxembourg. They are dependent upon regular contact with expert advisers from national governments and representative non-governmental organisations, in order to broaden the basis on which their proposals are prepared. The Commission has the tasks of initiating the broad outlines of Community policy, and defining the practical details of policies to be adopted, as well as implementing them. The initiation of common policies has always been regarded by the Commission as one of its most important duties. The Treaty of Rome indicated the direction in which policies were intended to go, but left a major responsibility with the Commission to initiate specific policies. Moreover, in 1972, the Paris Summit Conference of heads of government required the Commission to propose common policies in fields not strictly provided for in the treaties themselves, including environmental policies in pursuit of the Communities' overall objectives.

In pursuit of its duty to initiate common policies the Commission makes proposals to the Council of Ministers. Indeed, the Council is dependent on this, because the Council can only deliberate on the basis of Commission proposals. The Council consists of a representative of each member-government. Normally this representative is the Foreign Minister, but other ministers often attend, including those for agriculture, industry, transport, or whatever is appropriate. In addition, heads of government have since 1974 been meeting at least three times a year, and these meetings are known as European Councils. Presidency of the Council rotates between member-governments at six-monthly intervals, in a sequence determined alphabetically according to each country's name in its own language. Belgium begins the sequence which started a new cycle in 1982, and the UK comes at the end. The Council is the ultimate decision-making authority, with the power to adopt measures which then have legal force within the Communities. The relationship between Commission and Council is summed up in the adage 'the Commission proposes, the Council disposes'.

The Council has its own secretariat, and is advised by the Committee of Permanent Representatives, known by its French acronym as Coreper, whose members are the permanent official representatives, or ambassadors, of the national governments. In addition, the opinions of the European Parliament and the European Economic and Social Committee usually are sought by the Council before it can act upon a proposal from the Commission.

The European Parliament normally meets in Strasbourg, and was directly elected for the first time in 1979. It has 434 members, and is a fully integrated Community institution, in the sense that its members align themselves in cross-national political groups, and not in national groups. Members also sit on committees which correspond to the main divisions of the work of the Commission, and to its directorates-general. Commission proposals are considered in detail by the appropriate commitee, which makes a recommendation to the Parliament. On the basis of this recommendation, and any other representations received by members, Parliament will debate a proposal and adopt a resolution expressing its opinion. Parliament also has certain powers over the Community budget.

As a parallel process Commission proposals are also considered by the Economic and Social Committee, or in the case of ECSC matters, by a Consultative Committee. This consultation is mandatory on many issues. The Economic and Social Committee consists of 156 members, drawn from the member-states according to an agreed allocation. Members are nominated by national governments, but their membership is personal. They are not government spokesmen or representatives. Normally members have achieved prominence in some walk of life such as industry, trade union affairs, and academic, professional, or other specialist fields. They group themselves not by nationality or political party, but into three major groups, consisting of employers; workers' representatives, trades unionists, and so on; and those representing various other interests such as the professions. The full committee meets in Brussels ten times a year, and these meetings are preceded by meetings of the nine sections. Each section is responsible for considering matters concerning a particular aspect of the Commission's activities. When the committee is consulted, the appropriate section considers the proposal, and in doing so may set up a study group and seek the advice of appropriate independent experts on the subject in question. A recommendation is then made to the section which will prepare a draft opinion. This is then received and debated by the full committee, which then issues its opinion.

The opinion of the Economic and Social Committee, and the view of the European Parliament, are taken into account by the Council in deciding upon a Commission proposal. Both the committee and the Parliament may also issue opinions on their own initiative on matters of Community policy.

Having received proposals from the Commission, and the responses to all the various consultations, the Council can take action. Under the Treaty of Rome the Council resolutions may be adopted by a system of majority voting, known as a weighted majority, whereby a balance is created between the influence of the different member-states, so that the large ones cannot overrule the smaller members, but one or two members cannot prevent progress by impeding measures that command widespread support. In fact, most major measures are adopted on the basis of unanimous support, or at least with abstentions instead of votes against. Since the Luxembourg Compromise of 1966[4] the practice has been to allow a form of veto whereby individual member-states may block a measure in the Council if it claims that its vital national interests would be adversely affected.

Binding action can be taken by issuing regulations, directives, or decisions. Regulations are of general application, binding in their entirety, and directly applicable in all member-states. Directives are binding on member-states as to the result to be achieved, but the means by which this is done is left to the discretion of the competent national authorities, who would initiate any legislation necessary. An example of a proposed directive in the planning field on environmental impact assessment, is discussed below. Decisions may be addressed to a particular government, enterprise, or individual, and are binding in their entirety upon those to whom they are addressed. In addition, under the EEC Treaty, the Council may issue a recommendation or an opinion, neither of which is binding. A recommendation under the ECSC Treaty is, however, binding, and corresponds to an EEC directive.

Having outlined the main policy-making institutions, and indicated the means whereby they act in pursuit of their responsibilities under the treaties, it is now possible to consider the policies themselves in those aspects of the Commission's responsibilities which relate to planning. These may be grouped into policies concerned with environmental protection; financial provisions providing for infrastructure investment, such as the Regional Fund, or for skill-training, such as the Social Fund; and measures concerned with professional qualifications and practice. Research into processes of urbanisation and planning issues, as a basis for the further development of policies, has also been sponsored by the Commission.

ENVIRONMENT POLICY

The Communities' environment policy was initiated following the declaration of the Paris Summit of heads of government in October 1972. Responsibility for executing it now rests with directorate-general XI: Environment Consumer Protection and Nuclear Safety. Since there is no explicit call for an environment policy in the Treaty of Rome, in contrast to the requirement for common transport or agriculture policies, the legal basis for an environment policy has had to be found elsewhere, in the more general articles of the Treaty. Article 100, which confers general powers to act on harmonisation proposals, provides the legal basis for the majority of environmental measures. It states that:

> The Council shall, acting unanimously upon a proposal from the Commission, issue directives for the approximation of such provisions laid down by law, regulations or administrative action in Member States as directly affect the establishment or functioning of the common market.
>
> The Assembly and Economic and Social Committee shall be consulted in the case of directives whose implementation would, in one or more Member States, involve the amendment of legislation.[5]

The other general power sometimes invoked in the context of the environment policy is article 235, which confers a general power to initiate policies not otherwise provided for in the treaty. It states that:

If action by the Community should prove necessary to attain, in the course of the operation of the common market, one of the objectives of the Community and this Treaty has not provided the necessary powers, the Council shall, acting unanimously on a proposal from the Commission and after consulting the Assembly, take the appropriate measures.[6]

The impetus given towards an environment policy by the Paris Summit of 1972 enabled the Council to adopt the first Action Programme on the Environment, for the period 1973–6, in November 1973.[7] This action programme was renewed and updated by the Council in May 1977,[8] for the period 1977–81. A *Report on the State of the Environment* was issued in 1977[9] and in 1979,[10] and in May 1980 the commissioner responsible, Lorenzo Natalia, presented the president of the Council with a statement of progress made on the environment action programme, and on the work done to implement it.[11] A third action programme on the environment was proposed by the Commission in 1981, to cover the period 1982–6,[12] and this was adopted by the Environment Council in December 1982.

The initial concern was that, as a result of different material policies on product standards and emission control, disparities would arise creating distortions of competition. The first action programme was primarily concerned with remedying existing problems in specific industries.

The second action programme has four main elements. The first of these concerned the reduction of pollution. More progress was made towards reducing water pollution than atmospheric pollution, partly because of the more specific and tangible nature of water pollution. Standards were prepared for industrial emission or fuels used for air or land transport, and for waste management and disposal. The second involved active engagement in international co-operation to tackle environmental problems whose scale goes beyond the dimensions of the Community itself. Thirdly, a major area of work concerned the rational management of natural resources, including non-renewable resources, water, flora, fauna and the land itself. An important objective of this aspect of environmental work is to integrate the environmental dimension into economic growth – in order to orientate economic growth in a direction which will protect the environment and enhance the quality of life to be anticipated by future generations. The principle that prevention is better than a cure, recognising that it is insufficient for environmental policy merely to respond to problems already created by industrial development, underlies much of the programme. The fourth element consists of detailed studies and research into individual technical topics, and the promotion of environmental education and public awareness of environmental problems.

The third action programme is designed to continue to make progress in these directions. However, its significance for planners is greatly increased by the fact that it places considerable emphasis on physical land-use planning measures, and in much more explicit terms than previous statements have done. The principles on which it is based include the 'polluter pays' principle and that prevention is better than cure, and the principle that measures should be implemented at the appropriate geographical and political level. The first of these seeks to ensure that the cost of appropriate prevention or remedial measures should be borne by the enterprise responsible for a potential source of pollution. Prevention is emphasised as being preferable to remedial action both on grounds of cost and of sound environmental planning. Control of land use, and environmental assessment of land-use development proposals, is recognised as being a fundamental means of pursuing this principle. Finally, since land-use planning powers are for the most part held by local, provincial, or regional authorities in the member-states, as the foregoing chapters have demonstrated, the principle of the appropriate geographical and political level implies that any land-use planning measures adopted under the environment policy would be executed by these authorities, and not by the Commission or national governments directly.

The third action programme contains a range of proposals similar to those in the second programme, several of which are not of direct concern to land-use planners. The first part, on overall strategy, includes procedures to ensure that environmental data are taken into account in planning. A major element of this, the environmental assessment of projects, is discussed below, and similar measures for development plans and programmes are envisaged. The creation of an environment fund, initiated in the 1982 Community budget although with only a symbolic allocation, is also proposed. The second part is specifically concerned with continuing the programme of measures for individual pollutants. The third section is concerned with the protection and rational management of land, the environment and natural resources. Land is recognised as being a vital, scarce resource, and 'Physical planning is therefore one of the areas where a preventive environment policy is very necessary and very beneficial'.[13] Specific proposals include a call for closer co-ordination between environment policies and sectoral policies such as agriculture, transport and regional development, and for a system of ecological mapping. Studies are underway with a view to developing an ecological mapping system to present in a standardised manner data on the natural environment, as a basis for planning the use of land. It will aim to relate a scientific description of the characteristics of the land surface to its significance in terms of physical planning. When operational, it is intended to assist in the planning of specific development, particularly when Community investment under the regional, social, or agricultural programmes is planned. The final part of the programme carries forward the international element of the second programme. It recognises that many environmental problems extend beyond the boundaries of the Community.

The relationship between environmental policy and the economy is one that receives much attention. Harmonisation of regulations for control and authorisation of particular processes or activities is pursued not only in order to raise the general level of environmental protection, but also to overcome any distortions of trading conditions that would arise from widely differing policies being adopted in different member-states, with the consequent creation of pollution havens, where controls are weak. Different standards of emission, for instance, could affect the profitability of enterprises in different countries.

The relationship between environmental policy and employment must be considered, especially in times of economic recession, because it is frequently presumed that measures for environmental protection add to the costs of industry, reducing its profitability and capacity to sustain employment. The Commission argues strongly, however, that for a number of reasons the adoption of such measures has by no means a totally negative effect on employment, especially if harmonisation is achieved throughout the Community. The marginal cost of environmental protection measures may be slight, and the administration and technology of environmental protection itself creates jobs. Most importantly, several sectors of the economy such as agriculture, fisheries, tourism and recreation have a vested interest in the quality of the environment, and industrial productivity generally is improved by the reduction of environmentally induced health hazards such as respiratory illnesses.

As is clear from the foregoing accounts of the national planning systems in the EEC, a great variety exists both in the principles and procedures adopted and in the impact of planning controls in the various member-states. This is to be expected as most of the systems described were conceived well before the foundation of the EEC. The imposition of some degree of harmonisation by the EEC of land-use planning procedures is, therefore, clearly a highly complex and intricate task, involving extensive negotiation.

The planning measure which has progressed furthest is a proposal for a directive on the 'Assessment of the environmental effects of certain public and private projects'. Although the proposal does not use the term, it is in effect a proposal requiring a form of environmental impact assessment (EIA) to be undertaken prior to authorisation of major developments anywhere in the Community.

This proposal was published in June 1980[14] and has now been considered by the national governments, the European Parliament, and the Economic and Social Committee. In the light of responses received from these bodies and interested non-government organisations the Commission has published amendments to the proposal in March 1982.[15] It is now 'on the table' for adoption by the Council of Ministers, having first been discussed by the Council in June 1982. This proposal is the first major measure to be concerned with the authorisation of development by land-use planning procedures, and it would introduce for the first time a limited element of harmonisation into the Communities' variety of planning systems.

The essence of the proposal is that, when a developer seeks authorisation for a major development in one of the categories listed, he must supply the appropriate authority with an EIA dossier, alongside whatever form of planning application may be required. This dossier would indicate the anticipated impact, and the measures to be taken to minimise it. Developments where this would arise include a variety of chemical and other industries handling toxic materials, mineral extraction, nuclear installations; and infrastructure investment in motorways, airports, power stations, and so on. The developer may, of course, be a government department, a public agency, or a private developer. Consultations with appropriate agencies, and the general public, would take place on the basis of the EIA dossier before any decision is reached. By addressing issues at an early stage, and opening up the process to public scrutiny, it is hoped to achieve an improvement in the environmental consequences of major developments, greater public confidence in the decision-making process and quicker decisions. The potential for quicker decisions is most apparent in cases involving highly controversial proposals, such as those which, in the UK, have led to protracted and contentious public inquiries.

Some form of EIA procedure already exists in some members of the EEC, such as France and Ireland. Although the UK is not one of these, its incorporation into the British planning system would not change significantly the degree of control or process of determining whether to grant planning permission in the case of well-presented private sector proposals, especially if, as is likely, responsibility for considering EIA submissions is assigned to local planning authorities. It would, however, bring rather more changes to the procedure for considering government proposals, or proposals which are outside the scope of planning control at present, such as those relating to agricultural development. British planners tend to be concerned with economic or social consequences of proposed development, and these aspects together with consideration of consequences for the natural environment, could be expected to receive consideration in Britain. It is by no means certain that EIA would be operated in such close association with the statutory land-use planning system in all other member-states as it would be in Britain. If operated in a uniform manner throughout the EEC, with similar environmental quality criteria, it would go some way towards harmonising the approval procedure and conditions under which authorisation might be granted for a number of significant categories of development. However, it will be a long time before its effects, and the degree of harmonisation achieved, can be judged.

REGIONAL AND INDUSTRIAL INVESTMENT

The concept of setting up a European Regional Development Fund had been under consideration since the 1960s, but significant progress was made in association with Britain's negotiations to join the EEC at the Paris Summit of 1972. George Thomson, one of the first two British commissioners, held the regional policy portfolio and, in 1973, put forward proposals for a regional fund. Eventually the Regional Development Fund was set up, with effect from 1 January 1975.

The objective was to boost the economy of the less favoured regions of the Community, the peripheral areas, the economically underdeveloped regions, and the regions suffering from decline due to their industrial structure or outmoded infrastructure. It was hoped that the fund would reduce the disparity between the richer and poorer regions of the Community to a greater extent than could be achieved by national regional programmes alone. To this end, assistance from the fund is intended to be additional to the regional assistance offered by national governments, and should not be a substitute for it.

A quota system was established giving each member-state a fixed share of the fund. Under the quota the major beneficiaries have been the poorer members, namely, Italy (40 per cent), the UK (28 per cent) and Ireland (6 per cent). France also had a considerable share (15 per cent). These percentages were adjusted to make provision for Greece to receive 13 per cent. None of the nine previous members has suffered an absolute reduction in aid since the overall fund has increased. The UK quota became 23.8 per cent, the Italian 35.5 per cent, the French 13.6 per cent and the Irish 5.9 per cent.[16] In addition, a small percentage (5 per cent in 1980-2) of the total fund is non-quota. This means that it can be disbursed outside the confines of these percentage quotas, to counteract major problems of industrial decline, such as in shipbuilding or steel closure areas in the UK, and in the border area between Ireland and Northern Ireland.

The Commission has favoured a larger non-quota section in order to increase the flexibility of operation of the fund and, in 1981, proposed substantial changes.[17] These proposals involved allocating a quota only to the member-states with the greatest regional problems (Greece, Ireland, Italy and the UK), and giving assistance elsewhere through an enlarged non-quota section. It is far from certain that these proposals will find approval, however.

Aid from the main fund may be used for (1) investment in the production of industrial goods or services; or (2) for investment in infrastructure projects related to industrial or tourist development, and lately also for derelict land reclamation, or social infrastructure in urban areas; and (3) for farming in hilly and remote areas. Projects are eligible if they are located in areas included within an existing national programme of regional development. Formal application must be made by the appropriate national government, and grants go to the government who choose whether to pass it on as a direct payment to whoever is responsible for a particular project. Proposals for assistance by the fund often relate to local authority development projects, and planning departments are frequently responsible for assembling packages of proposals which form the basis of formal applications by the national government. To be successful a proposal must fall within an area designated for regional assistance, and fulfil the Commission's own criteria concerning Community goals of free trade, harmonisation and reduction of disparities in regional economic development.

Finally, it should be noted that the regional fund is not permanently established. The initial fund, set up in 1975, was for a three-year period. Since then it has been renewed, but not placed officially on a permanent footing, although the Commission has proposed that it should be. On the other hand, it is improbable that a decision would be made to terminate the fund.

Although less directly related to planning, certain other forms of Community financial aid should be mentioned because they have a direct bearing on financing development, or improving employment prospects for residents in areas of high unemployment. The European Social Fund is used to assist in training or retraining to give new vocational skills, especially where there are large numbers of unemployed, or young people entering the labour market and there is a need to teach skills appropriate to new industry, replacing obsolete skills where necessary. Use of this fund may be made in association with policies of economic regeneration such as those adopted by a number of local planning authorities in the areas of Britain affected by the structural decline of industry. Another source of finance for economic development, in the form of loans for major capital investment, is the European Investment Bank. Its funds are to be primarily devoted to industry, energy, or infrastructure investments, especially where they might assist in overcoming regional problems, or support industrial modernisation, or projects of a multinational nature. In contrast with the regional fund, both the social fund and the European Investment Bank are permanent, and are provided for in the Treaty of Rome. Provision similar to the social fund, exists within the framework of the

European Coal and Steel Community, in the form of a Retraining and Settlement Fund to assist with the problems to be found in areas of declining coal or steel production.

TRANSPORT

The need for a transport policy is explicitly referred to in the Treaty of Rome, because it is regarded as one of the principal means of removing barriers to trade between member-states. A common transport policy has, therefore, been developed in order to meet the objectives of the treaty. Until recently it has concentrated on the activities of public and private transport operators, with the objective of removing restrictive regulations which would inhibit transport operators from developing Community-wide services, while at the same time maintaining and harmonising essential safety standards. Progress has not been entirely satisfactory, although the volume of transport movement between member-states has of course increased greatly over the years. In a Commission memorandum on the role of the Community in the development of a transport infrastructure adopted in November 1979[18] development of a common transport policy is outlined. While national transport networks have been extensively improved, particularly as the memorandum notes, with a disproportionate priority for the development of roads and motorways, infrastructure provision for intra-Community transport has not been improved to the same extent. It is proposed that studies be made with a view to preparing a Community policy for transport infrastructure. This would be based upon existing national networks, but specifically designed to meet the increased demands for intra-Community movement. Major existing links and possible improvements to the road, rail and waterway networks are indicated. These proposals are by no means confined to cross-border transport links. Investment on routes well removed from borders, but which are a part of the network of movement for Community trade, are also envisaged. A major example of this type of project is the study announced by British Rail in 1980 for a rail tunnel underneath London linking the main rail links to the north of England with the links to the Channel ports. As the Commission itself notes, the orientation of the common transport policy in the directions outlined has major implications for land-use planning. Not much progress has yet been made towards formulation of a common policy, yet however.

THE PLANNING PROFESSION

The Treaty of Rome provides for free movement of labour, and freedom to establish professional practices. Provisions for the free movement of labour allow people to seek paid employment, and accept employment without first obtaining a work permit, in any member-state, under conditions affording equal status with nationals of the country in most important respects. These provisions would apply to planners seeking salaried employment, although the wide variation in planning processes and legislation means that very few people have taken advantage of them.

Exercising the right to establish private planning practices other than in one's own country is similarly limited by the variety of planning systems, but the situation is further complicated by the varying extent to which planning is recognised as a separate profession, distinct from architecture, civil engineering, or surveying. There is no present intention to prepare a directive on the recognition of qualifications or professional titles for planning, but such a directive is proposed for activities pursued under the professional title of architect. In some countries this would include tasks which are performed elsewhere by planners who were not architects. Architecture is one of the professions which is regulated by law in several countries, whereas town planning generally is not. In the view of the Commission there do not appear to be any restrictions on the right of establishment for planners requiring Community intervention. However, there is also no basis agreed by the Commission for mutual recognition of planning qualifications or protection of professional titles such as that of chartered town planner, protected in the UK by the Royal Town Planning Institute.

FUTURE DIRECTIONS

The relationship between policies of the EEC and the activities of local authorities responsible for town and country planning is deepening, as the EEC develops its environmental policies and investment in industry and infrastructure under the regional, social and transport programmes. The Commission recognises, however, that the differences in environmental quality and economic prosperity to be found within regions, and within major cities, are often greater than the differences to be found between different regions. Furthermore, it is recognised that the manifestations of urban change and urban stress are similar in different parts of the community.

As Lorenzo Natali, vice-president of the Commission, indicated in a speech to the conference on urban problems in the Community, held in Liverpool in November 1979, the Commission is increasingly concerned with both economic and environmental issues in the urban context. Consequently, one can conclude that issues and policies of concern to planners are receiving increasing attention in the institutions of the EEC. This is true whether one considers the action programme on the environment, the regional and social policies, or the common transport policy. As further possible courses of Community action are suggested, the implications for planning authorities and practitioners will become greater. There is, therefore, an increasing need for planners to be informed about initiatives being made by the Community, and for professional institutes and local planning authorities to establish a dialogue with those responsible for Community policy on matters of mutual concern. In this way, Community proposals are more likely to take account of local factors in an acceptable way, and implementation of such proposals by planners and planning authorities is less likely to present problems.

NOTES: CHAPTER 12

1 See Swann, 1978, p. 19.
2 See Whittaker, 1979, p. 1. The author is chairman of the British sections of the International Union of Local Authorities, and Council of European Municipalities.
3 See Noel, 1979, p. 65.
4 See, for example, explanation in Swann, 1978, pp. 65-6.
5 *Encyclopedia of European Community Law*, Vol. BII, European Community Treaties, London, Sweet & Maxwell, 1982, B10086.
6 ibid, p. B10166.
7 *Official Journal*, C112, 12 December 1973.
8 *Official Journal*, C139, 13 June 1977.
9 *First Report on the State of the Environment*, Brussels, CEC, 1977.
10 *Second Report on the State of the Environment*, Brussels, CEC, 1979.
11 *Progress Made in Connection with the Action Programme on the Environment*, Brussels, COM(80) 222 Final, 7 May 1980.
12 *Draft Action Programme on the Environment, 1982-6*, Brussels, COM (81) 626 Final, 4 November 1981.
13 ibid, s. 26.
14 *Draft Directive concerning the Assessment of the Environmental Effects of Certain Public and Private Projects*, Brussels, COM (80) 313 Final, 16 June 1980.
15 See *Proposal to Amend the Proposal for a Council Directive Concerning the Assessment of the Environmental Effects of Certain Public and Private Projects*, Brussels, COM (82) 158 Final, 31 March 1982.
16 EEC Regulation 3325/80.
17 *Official Journal*, C336, 23 December 1981.
18 *A Transport Network for Europe*, Bulletin Supplement 8/79, Brussels, CEC, 7 November 1979.

Chapter 13

Some Links and Comparisons

R. H. Williams

The reasons for having a system of town and country planning, the role and function it is expected to perform, and the quality of the advice and decisions that emerge from the planning system, are frequently a matter for debate and political dispute within individual countries, but it is sometimes less apparent to individual practitioners that many of these issues are faced not only in one's own country, but in several others as well. It would be possible to anticipate all the points of comparative interest from the foregoing chapters, and no attempt is made to do so. Still less is it intended to make any suggestion about which national systems of planning are better, which features deserve emulation in other countries, or which lessons may be learned by whom. Readers will form their own judgement, and will identify for themselves lines of inquiry which may merit further study.

In general terms, I share Cherry's preference for seeing planning as being 'culturally derived, and developing within political and institutional boundaries'.[1] However, the various national planning systems of the EEC countries have clearly not developed in isolation from one another, although each shows characteristics derived from the history and institutions of its own country. A number of common themes do emerge, and an attempt is made here to identify some points of interest relating to the evolution and current operation of the planning systems, to the ideas and theories behind them, to their administration and to the planners who operate them. Finally, some aspects and applications of comparative analysis are discussed.

These accounts of planning systems have not, in general, looked back before the postwar period. It is striking, however, to note how frequently the period of hostilities can be seen to be a major stimulus for the development of town planning. Obviously reconstruction was a major priority, with a great deal of town planning work to be done in a short period, often with a shortage of suitably qualified people. This period has had a major influence on the development of planning in several countries. It is more surprising, perhaps, to observe how the period of hostilities itself produced new planning legislation or agencies. One may cite Italy, where the law of 1942 is still the basic planning legislation, or the Netherlands, where the National Physical Planning Agency, established in 1941, still plays a major role in the relatively highly developed system of planning at the national scale, as well as advising municipal and provincial authorities, disseminating research, and so on. A further case to note is Belgium, whose government-in-exile – in both 1915 in Le Havre, and 1940 in London – resolved to enact town and country planning legislation, seeing this as a necessary element in orderly reconstruction.

This association between the development of proposals for planning systems and the destruction of war seems to be too widespread to be pure coincidence. It suggests that the general desire to consider reconstruction and the creation of a better environment in which to rebuild communities in the postwar era inspired in many people in different parts of Europe the specific desire to create an effective system of town and country planning. Thus, a link may be established between the two themes binding together this book, namely, town and country planning and the EEC, since the latter owes its origin to the resolve of its founders to construct a safer Europe and remove the risk of further European wars.

REORIENTATION OF PLANNING

Conceptual thinking about the nature and methods of planning has, of course, proceeded a long way since the Second World War. One particular aspect of this, noted in several countries at a similar time, is the shift away from the perception of planning as a purely architectural or technocratic and politically neutral activity, to the acknowledgement of a more social policy or programme-oriented approach, sometimes described as a more humane style of planning, or a more explicitly politicised approach to policy-making. A shift of emphasis of this sort seems to have been very widespread in Europe since the late 1960s and during the 1970s.

In several cases this was associated with reforms of the planning system, or of local government and powers of local planning authorities. This shift of emphasis was stimulated in the UK by reforms of both of these during the period 1968–75, the associated influx of new personnel, and the need to rethink policies. In Denmark this shift of emphasis can also be linked both to municipal reforms and to reforms of the planning system, though at a slightly later period in the 1970s. A number of other countries including Belgium and Germany underwent some local government reforms during this period, but in most cases the formal procedures of planning remained as they had been. The shift of emphasis is linked in Germany with the great rise in oil prices in 1973, and the realisation that continuous economic growth could no longer be taken for granted.

Economic circumstances in all countries underwent changes during this time. Greater competition for reduced resources meant that development proposals were subject to greater scrutiny, implementation was not always assured and greater emphasis was placed on conservation in all its forms. The wisdom of the scale of development and redevelopment in the 1960s was increasingly called into question. Above all, there was a growing awareness among the general public, and among political parties, of the social and economic as well as physical costs and benefits that different forms of development and land-use policies had for different groups in society.

There were a number of responses to these changes. The introduction and general acceptance in most of Europe of some form of public participation in the planning process was one of the first and most fundamental responses. Changes were made to legislation in several countries to accommodate greater public participation in the planning process, the UK being one of the earliest in 1968. Public participation has become generally accepted as an essential element of the planning process, whether or not it has been introduced by legislative change, thus implicitly acknowledging that planning decisions are in the realm of the body politic, and are not to be regarded as purely technical matters for the professionals alone.

Other innovations in planning occurred during this period. Greater emphasis on conservation of natural environments, on linking proposals with the resources for implementation, on linking economic development with physical planning, and less radical and unorthodox changes in urban forms and style of housing featured among the responses to changing economic circumstances, and changing political attitudes towards planning, which occurred over most of Europe during the late 1960s and 1970s.

More recently wider environmental issues have come to the fore in politics with the rise since the late 1970s of environmentalist, or Green, political parties. This has been especially significant in Germany and France, where they have achieved some electoral impact, but this movement has become a factor in the politics of planning in several other countries now, including Belgium, the Netherlands and, through pressure-group activities, in the UK.

DEVELOPMENT PLANS AND CONTROL OF DEVELOPMENT

Several countries have some form of strategic plan at region, province, or county level, usually acting as a context for more detailed local plans and development control. Some countries begin a step before this, and attempt some form of national planning. Luxembourg naturally adopts this scale of planning but, of course, this corresponds to a local authority or medium-sized city in scale and population. Other countries that adopt the principle of planning at the national level including Belgium, Denmark, France, Greece and the Netherlands. Often this takes the form of guidelines or objectives as in France, or may be linked to economic development as in Denmark or Greece. The Netherlands has the most fully developed system of physical planning at national level among the larger countries, but this appears as an exception to the general rule.

Elsewhere the extent and significance in practice of physical planning at the national scale has been very limited, even where it is explicitly required by law as in Belgium. Perhaps one could conclude that general land-use plans are an inherently inappropriate form of policy-making at national level in countries with the complexity of those studied here. Sector plans, such as those for roads in Federal Germany, appear to be a more effective form of planning policy-making at the national level.

Planning at regional scale is more widespread, and exists in some form in most countries, although in some cases, such as Britain, it is no longer pursued extensively. Early postwar British regional planning, particularly the Abercrombie plan for the London region, provided a stimulus for other countries such as Denmark. Entry into the EEC, and the opportunity to benefit from the Regional Development Fund, is noted as having stimulated thinking about regional policy and planning in Greece. In the context of regional planning the concept of a regional growth pole occurs in a number of countries. Belgium, the Netherlands and Greece, for instance, have at various times adopted policies of regional development based on Perroux's growth-pole theories[2] and the concept is implicit in some new town locations in Britain.

An upper, or strategic level of local authority planning distinct from regional planning can often be identified. The UK, however, is unorthodox in that this form of plan, the structure plan, is not only the highest level of formal plan-making, but also the only mandatory form of plan, and the only scale at which a complete coverage is envisaged. Structure plans are the only part of the British development plan system designed to cover the country's total land area, and to be approved by the appropriate secretary of state. An equivalent level of plan-making exists in several other countries, such as Germany or the Netherlands with the *streekplan*, but in all of these cases its role is advisory, as a context for a mandatory form of local plan or building plan.

At the local level several countries have two types of plan, one at the level of a whole municipality or community, and one for a very specific area or precinct within which development is assumed or proposed to take place. Germany, for instance, has the *Flächennutzungsplan* (land-use plan) at the former level, and *Bebauungsplan* (building plan) at the latter. The French SDAU and POS, Dutch *Structuurplan* and *Bestemmingsplan*, or Belgian local plans, have a similar relationship. In each case it is the lower, or more detailed, level of plan which is mandatory in appropriate circumstances, but without any suggestion that they should cover all the land surface. Plans of this type normally are prepared for areas which are known to be intended for development, and they have the effect of indicating precisely what form of development would receive permission. A considerable degree of architectural detail may be incorporated, and a proposal to build that is consistent with the plan may in several countries receive permission, without any discretionary power for the competent authority to decide otherwise. This principle is at its most explicit in the Netherlands, where the concept of *Rechtstaat*, offering legal certainty to its citizens, is reflected in the planning system. Nevertheless, several of these systems can achieve a great degree of flexibility in practice. Some, such as France, have a more discretionary system operating in those areas not covered by a POS, while in the Netherlands extensive use is made of a procedure for by-passing the rigidity inherent in the formal system.

The UK does not have the formal system of two tiers of local plan referred to above, but the equivalent could occur in the sense that a district plan may correspond to the municipal or community plan, and an action area may correspond to a precinct or building plan. However, the similarity does not extend very far, since the degree of detail, extent of national coverage and status in relation to the control of development are different in Britain. Britain and Ireland both adopt the principle that discretion rests with the competent authority to grant or withhold permission, whether or not a proposal is consistent with any local plan. Obviously decisions generally reflect the policies of the plan, but there is nevertheless a greater degree of flexibility, or uncertainty, inherent in both these systems than there is, theoretically at least, in several of the systems operating in mainland Europe. Both Britain and Ireland have, as a consequence of this loose relationship between formal plans and permission to build, elaborate procedures for appeals against decisions to be heard. In the case of Ireland this includes the creation of an independent Planning Board, and provision of a universal right of appeal to it. The operation of this institution is potentially of great interest elsewhere.

These discretionary systems contrast with the certainty attaching to a system whereby planning or building permission is known to be forthcoming for proposals in accordance with the local plan. It has been suggested in the UK that the greater certainty for the applicant or developer offered by the planning regime in enterprise zones would have a beneficial effect in stimulating development. This may possibly be interpreted as a small move in the direction of continental planning systems.

In addition to the various forms of planning or building permission required under planning legislation, there are frequently other forms of authorisation that may need to be sought for different types of development. In Luxembourg, for instance, it is possible to envisage a total of seven separate authorisations being required for one project. A number of countries now possess procedures for the environmental appraisal of projects. Provision exists, for instance, in Ireland; in France, within nature-protection legislation; and in Germany, within emission-control legislation. The Commission of the European Communities published in 1980 a proposal for a directive on the assessment of the environmental effects of certain major projects.[3] Devising a procedure which can be adopted for use within the variety of existing planning and control systems can be seen to be a highly complex task, and at the time of writing the directive has not reached the stage of being adopted, although it has been the subject of very widespread consultations, as a result of which a number of refinements to the text have been drafted. In the light of the great variety of town and country planning systems, both in terms of scope and extent of powers, and in terms of effectiveness, it will be recognised that any specific proposals the Commission may bring forward under the third action programme on the environment[4] for rational land use, or the assessment of plans and programmes, will have to be capable of adjusting to this highly complex pattern.

In this review consideration has been given first to the national and then regional scales of planning, followed by more local and detailed scales. This top-down approach is implicit, at least in theory, in most systems of planning, although sometimes more progress has been made in preparation of the more local plans. Denmark, however, is notable for adopting explicitly the concept of building up from the most local scale of planning to the larger scales, and Italy has implicitly accepted this mode of procedure until recent years.

POLICIES AND PROGRAMMES

Frequently, similar policies, forms of designation for a particular purpose, or programmes of action appear in different countries, in such contexts as new towns, conservation, or urban renewal. For this reason, policies of this type may offer the prospect of comparative study, or the

Some Links and Comparisons   165

transfer of experience or technique from one country to another. In the case of new towns, for instance, it is not unusual for the British programme of new towns initiated in the 1940s to be seen as a stimulus for similar developments elsewhere, although many of the distinctive characteristics of new towns in Britain are not always present in other examples. The basic concept in Britain was that a new town should be largely self-sufficient in employment, shopping and community facilities. Adherence to this has not remained as strict as was originally conceived, but this principle has distinguished new towns from overspill housing projects in Britain. A number of new town developments elsewhere, such as Køge in Denmark or Nord-West Stadt near Frankfurt in Germany, have not followed this objective of being so clearly self-contained, particularly with regard to employment and journey to work. The French have also developed a new town programme, and in doing so have adopted another distinctive characteristic of British new towns, namely, a separate *ad hoc* planning administration, in the form of a development corporation. The basic concept of a special agency to implement the new town is common to both countries, but in France the board of the corporation includes a substantial number of local councillors. This should overcome the criticism sometimes levelled at British new town development corporations, that they are insufficiently democratic and accountable to the local residents.

A wide variety of conservation measures have been adopted in several countries, both for protection of nature and of the built environment. There are many sources of potential confusion in their terminology used for different forms of conservation, however, particularly with terms applied to the natural environment in one case and building conservation in another. For instance, in Ireland an area of special amenity refers to the natural environment, and conservation orders relate to the protection of species of wildlife; whereas in Britain a conservation area is a designation defining an area of a town or village where special measures to preserve the character of the built environment are adopted.

The emphasis in conservation or preservation policies varies considerably. At one extreme, concern in Greece is primarily with preservation of ancient remains, and conservation of the urban environment of the modern era is less emphasised. Elsewhere conservation has taken the form of reconstruction of buildings destroyed, especially in war-damaged German city centres. In Britain a clear distinction in principle is made between conservation of a working environment, and preservation of individual buildings of particular historic or architectural interest.

One policy that has been widely adopted in Europe, and often found in town centres with a conservation interest as well as in newly developed shopping streets, is that of pedestrianisation. An important early example is the Lijnbaan in Rotterdam. Also many German cities have had pedestrian zones for many years, which have attracted great interest as models of successful planning policies. Similar schemes have subsequently become a feature of urban policy widely adopted elsewhere in Europe.

POLITICS AND ADMINISTRATION OF PLANNING

It is evident from the emphasis placed repeatedly on government structures in the foregoing accounts of national planning systems, that administrative and political responsibility for planning, and the governmental structures established to implement the system, have a major influence on the style of planning practice within each country, and that they in turn reflect the legal and constitutional framework of that country. Consequently, the principal features of the governmental structure must be understood as a basis for understanding a country's planning system. Certain important variables stand out. The degree of central government control, and extent of local authority independence in planning matters, is very important. The structure of local government and any reorganisation is also important, especially in the British case, as is the distribution of professional planners between central government, local government and private practice. Public participation has become a significant feature of planning in many countries in the period alongside the move from a technocratic to social perception of the nature of planning discussed above. Provision for participation is a noticeable feature of legislation during this period in most countries other than France, but the extent of true participation is often very limited, especially on bigger issues and larger-scale plans, and in Belgium a feeling that it was somewhat futile was noted. Belgians are not necessarily alone in this regard.

The extent of centralised control of the planning system tends to be greatest in countries like France, Luxembourg as a result of its small size, and Greece. The French network of local branches of central government exercises close direction of the planning system, through its role in preparing and approving development plans, although this may change under Mitterrand's presidency. Many of its responsibilities are carried out by local authorities in Britain, for instance. Greek administration has followed the French model, and also shows a relatively high degree of central control. Central direction of the traditional French type, as an inherent feature of the administrative structure of the country, must be distinguished from the national guidelines for planning policy established by central government agencies as is the case in the Netherlands.

Some devolution of power from central government to regional, provincial, or local authorities has taken place in recent years in several countries, either specifically in relation to town or regional planning, or as part of a more general decentralisation of power.

Devolution of planning powers to local authorities has happened in Britain, and to a larger extent in Denmark and Ireland. Denmark has adopted the principle that the initiative in policy-making rests at the local level. In Britain the 1968 Act loosened some of the controls over local authorities by adopting the principle that only structure plans, rather than all development plans, need to be approved by central government. Further changes in this direction in the planning system have happened since, after local government reorganisation in 1974, and especially under the present Conservative government. In Ireland the county managers and the planning board operate without needing to seek government approval for individual decisions.

A more general devolution of power from the British government to Scotland, Wales and the English regions was proposed in the 1970s, but was not implemented. However, general devolution of powers has occurred in a number of countries, and this has significant implications for the operation of the planning system in the countries concerned. Italy has undergone a major redistribution of legislative and executive power to the regions since 1970.[5] France has always had a highly centralised administrative structure, but some devolution of planning powers took place in the late 1970s, and this trend is likely to be assisted by the policy of President Mitterrand to diminish the extent of control by central government over local affairs. Belgium's constitution was based on the French model of a centralised state. However, devolution of power over domestic policy including planning to the three regions of Flanders, Wallonia and Brussels is now well underway.

In contrast to these examples of devolution of power from central government, the federal government of Germany initially had no direct planning powers in the early years of the republic. It was only in 1960, after detailed discussion of whether it would be constitutional, that any federal legislation on planning was passed, and executive power for planning rests largely with state and local authorities. The states also retain legislative power, and therefore some variations in planning procedures remain in spite of the unifying influence of federal legislation.

Apart from the German case, which is explained by the constitutional checks and balances deemed necessary when the federal republic was set up after the war, it is generally the case that planning has been regarded as a candidate for devolution of political or administrative control, either as part of a general policy of giving powers to regional authorities as in Italy and Belgium, or as a measure in its own right, as in Denmark and Britain. In the former case, especially in countries such as Italy where the regions have been given new legislative power, some regional variations in planning procedures as well as policies can be expected to emerge.

In addition to the widespread desire on the part of local communities to have greater control of their local affairs, which underlies many of the programmes of devolution, there may be some factors specific to town and country planning, encouraging its devolution. The greater expertise now available to local authorities, the greater public and political interest in local planning issues since public participation became widespread, and recognition that physical planning is more feasible at local and regional levels may all be factors underlying this trend.

THE PLANNERS

The evolution of planning from its origins in architectural and civil engineering practice is reflected in the organisation and educational qualifications of the practitioners of planning, although there is quite a variety shown in the timescales and extent of the emergence of a distinct, identifiable planning profession. In Germany, the Netherlands and Belgium architectural traditions and education have had a major influence over planning, although in all three cases there is evidence of a distinct trend towards the social science based education of planners, separately from architects, in some universities. Engineering is represented more strongly alongside architecture in France, as was formerly the case in Britain. In Italy planning and architectural qualifications are still largely interchangeable. In contrast, they have moved apart in Britain, where planning education has drawn increasingly on strong traditions of applied geographical and social science teaching. Indeed, graduates in the latter disciplines have for some years formed a major proportion of recruits to the planning profession in Britain, and it is possible that a similar trend may become evident in other countries as planning becomes increasingly a public sector occupation.

In Britain and Ireland the majority of professional planners are employed as local government officials, and private practice occupies a much smaller role than it does in most other countries, but there are some indications in countries like France, Germany and the Netherlands of a converging trend. No country has legislated for a protected title for the profession of planning as distinct from architecture, and in many cases it is difficult to identify or define clearly the body of people who may be called planners. Many professional associations and institutes exist, but none of those that represent planners as a distinct professional group compare in size of membership or influence over education and qualifications with the British Royal Town Planning Institute. Some, such as the Irish Planning Institute, or the Flemish Institute in Belgium, are of very recent origin.

One role that the British RTPI plays is that of an independent source of expert professional advice to government departments or politicians concerned with planning issues or proposals for new legislation. A somewhat similar role is performed in other countries by *ad hoc* or permanent committees of experts, whose advice is sought by government at a central, regional, or possibly local level. Such bodies exist in Greece, in France and the Netherlands as government-sponsored agencies, and Luxembourg and Belgium as national committees of experts. In Luxembourg it has the title of Economic and Social Council, and its brief is wider than planning issues alone. It is this type of institution that formed the model for the Economic and Social Committee of the European Communities.

CROSS-NATIONAL STUDIES

It is beyond the scope of this book to develop a systematic cross-national comparative analysis of the planning systems described above, although a comparative approach is likely to be adopted implicitly whenever one reads about a planning system other than one's own. The rationale, objectives and approaches to comparative analysis are discussed elsewhere.[6] Two principal objectives of comparative study of planning are widely recognised. One concerns the practical benefits for education and practice of studying experience outside one's own national context. The other concerns stimulus to the development of theory.[7]

Here it is intended only to offer some initial comments based on the observation that, within the variety of planning systems in the EEC, some common threads may be identified, and to consider the implications for cross-national aspects of policy-making, since this comes to the fore within the EEC with its supranational powers of policy-making in a way which does not in other international contexts.

First and foremost, it is necessary to define what 'planning' is in each country, as White emphasises.[8] The foregoing accounts of ten planning systems represent ten personal answers to this question, by planners native to the country they are describing. Differing emphases are placed by these authors on such themes as the statutory basis of the planning system; the ideological and theoretical concepts represented in it; operational practice; the institutions of government or other agencies with planning responsibilities; and the historic development of the planning system.

Another major comparative dimension is that of the scope or range of activities encompassed by the term planning. Planning systems are all concerned with allocation of physical land use, but scale of operation of the principal land-use policy instruments varies. The extent to which such themes as economic planning, transport planning, environmental protection, housing renewal, or industrial promotion are integrated with the core of land-use planning, and perceived as being within the realm of interest of planners and planning authorities, is however subject to considerable variety.

Hitherto, as a review of physical planning in Europe and the role of international institutions in planning, prepared by the Dutch government observes,[9] physical planning still takes place largely within national frontiers, and this variety within planning systems has not therefore been of great practical significance. Now, however, suggestions have been discussed in a Council of Europe seminar on the key role of physical planning in the protection and management of the environment and natural resources for a regional planning strategy to operate at the European scale.[10] Also, as Chapter 12 shows, several aspects of EEC policy have increasingly strong implications for physical planning.

Moves towards harmonisation of planning procedures, and the development of common forms of control of land use, such as the proposal for an EEC directive on environmental assessment of projects, will all demand prior understanding of the variety that exists among the ten planning systems, as well as of the variety of perceptions of the scope of planning, and of its terminology.

Comparative study of planning is of practical value not only for these reasons, but also for a much older reason. Interest in learning from the planning experience of other nations has been longstanding. For instance, the example of German planning, studied by Thomas Horsfall in the early years of the century,[11] was a major influence in Britain in the period leading up to the Housing and Town Planning Act 1909.

In recent years, there has been considerable interest in the potential value of learning from other European countries, and in the question of transfer. The complexity of planning procedures, and the intensity of the relationship between planning and its cultural, political and institutional context, is brought home when attempting to transfer any but the most straightforward of planning concepts and policies.

For instance, the concept of pedestrianisation of shopping streets is an example of a planning policy that is susceptible to transfer and application elsewhere without detailed analysis of the different planning and legal procedures in the countries under consideration. In complete contrast, urban renewal and redevelopment or improvement of older housing areas is a topic of very widespread importance, where many benefits may be obtained by studying experience elsewhere. For this to be possible, however, a deep understanding of the procedures and policies in the two or more countries being studied is necessary. Davies has tackled the question of transfer in a study of this and other topics

Some Links and Comparisons 171

structured in a way designed explicitly to facilitate transfer and implementation of new ideas, and his findings confirm the complexity of this objective.[12]

The planning of urban and regional development, and control of land use is regarded as a necessary task throughout Europe, and this volume has demonstrated the rich variety of procedures and practices that have been devised to accomplish this task. There are, no doubt, many more comparisons and contrasts that could be identified, and many lines of inquiry that would justify further investigation. It is not possible to do more than identify some links and comparisons here, and many themes remain to be identified and developed either with a view to cross-fertilisation of ideas in planning or in the context of any programmes of the Commission of the European Communities that relate to the activity of town and country planning in Europe.

NOTES: CHAPTER 13

1 See Cherry, 1982, p. 133.
2 See Perroux, 1964; also M. J. Moseley, *Growth Centres in Spatial Planning*, Oxford, Pergamon, 1974, pp. 3–7.
3 *Draft Directive concerning the Assessment of the Environmental Effects of Certain Public and Private Projects*, Brussels, COM (80) 313 Final, 1980.
4 *Draft Action Programme on the Environment, 1982-6*, Brussels, COM (81) 626 Final, 1980.
5 See Scattoni and Williams, 1978, pp. 38–40.
6 See, for example, Faludi and Hamnett, 1975; Masser, 1981a, 1981b; and White, 1978.
7 See Masser, 1981b, p, 3.
8 See White, 1978, p. 4.
9 See National Physical Planning Agency (RPD), 1979.
10 See Kunzmann, 1981.
11 See T. Horsfall, *The Improvement of the Dwellings and Surroundings of People. The Example of Germany*, Manchester, Manchester University Press, 1904; discussed in Cherry, 1974b, pp. 26–8.
12 See H. W. E. Davies, *International Transfer and the Inner City*, Reading, School of Planning Studies, University of Reading, 1980.

Bibliography

A NOTE ON FURTHER READING

The extent of material available for further reading varies greatly between the different countries, and the different themes introduced in this book. In several cases very little material is available specifically devoted to the planning system of that particular country, and the main source of further reading is therefore the appropriate chapters of other collections of papers treating several separate European countries. There is considerable variation in the selection of countries described in such works, however. Another source of further material is to be found in books which seek to make a synoptic overview of urban or regional development in Europe, or which are devoted to a comparative analysis of two countries or themes. A separate body of literature exists on the institutions and policies of the European Communities, within which references to town and country planning tend to be rather indirect. Finally, there are works of a theoretical nature relating to planning principles, or the comparative analysis of planning systems and policies.

Therefore, a classified bibliography with brief annotation where this is necessary is the approach adopted. The classification is based broadly on the groups of topics referred to above, but since there is always an arbitrary element in any *ad hoc* classification scheme, these lists have been grouped at the end of the book rather than being assigned to individual chapters, to facilitate reference to different lists.

(A) GENERAL: STUDIES OF EUROPE AS A WHOLE, OR OF SEVERAL EUROPEAN COUNTRIES

Best, R. H. (1979), 'Land use structure and change in the EEC', *Town Planning Review*, vol. 50, no. 4, pp. 395-411.

Blacksell, M. (1981), *Post War Europe* (London: Hutchinson).
A geographer's analysis of the supranational political institutions of Europe.

Burtenshaw, E., Bateman, M., and Ashworth, G. J. (1981), *The City in Western Europe* (London: Wiley).

Clout, H. D. (ed.) (1976), *The Regional Problem in Western Europe* (Cambridge: Cambridge University Press).
Includes Italy, France, Belgium, the Netherlands, Germany, Denmark and Ireland.

Clout, H. D. (ed.) (1981), *Regional Development in Western Europe*, 2nd edn (London: Wiley).
Includes EEC of nine countries, plus others in Western Europe.

Garner, J. F. (ed.) (1975), *Planning Law in Western Europe* (Rotterdam: North-Holland Publishing Co.).
Includes all except Greece, Ireland and Luxembourg from the EEC, plus six other Western countries, including Spain.
Hall, P. G. (ed.) (1977), *Europe 2000* (London: European Cultural Foundation).
Synthesises deliberations of international groups of experts.
Hall, P. G., and Hay, D. (1980), *Growth Centres in the European Urban System* (London: Heinemann).
Analysis conducted for International Institute for Applied Systems Analysis of Urban Centres in Western Europe.
Hansen, N. M. (ed.) (1974), *Public Policy and Regional Economic Development. The Experience of Nine Western Countries* (Cambridge, Mass.: Ballinger).
Refers to France, the UK, the Federal Republic of Germany, Italy and the Netherlands in EEC, plus Spain, Sweden, the USA and Canada.
Hayward, D. J., and Narkiewicz, O. (eds) (1978), *Planning in Europe* (London: Croom Helm).
Political analysis.
Hudson, R., and Lewis, J. R. (eds) (1982), *Regional Planning in Europe*, London Papers in Regional Science No. 11 (London: Pion).
Conference papers including Ireland, SE England, the Netherlands, Italy, Denmark, France, plus Spain.
Krejci, J. (1982), *Europe in Maps and Tables* (London: UACES).
Kunzmann, K. R. (1981), 'The protection and rational management of Europe's environment and national resources – the key role of physical planning', paper presented to EEC/Council of Europe Conference, Strasbourg, 17–18 December, 1981.
Kunzmann, K. R. (1982), 'The European regional planning concept', *Ekistics*, no. 294 (May/June), pp. 217–22.
Lee, R., and Ogden, P. (eds) (1976), *Economy and Society in the EEC* (Farnborough: Saxon House).
Papers on economic, regional, environmental, transport and other planning themes.
McKay, D. H. (ed.) (1982), *Planning and Politics in Western Europe* (London: Macmillan).
Political analysis of planning in France, the UK, the Federal Republic of Germany, the Netherlands and Denmark.
Minshull, G. N. (1978), *The New Europe – An Economic Geography of the EEC* (London: Hodder & Stoughton).
Sectoral and area analysis.
National Physical Planning Agency (RPD) (1979), *Physical Planning Developments in Europe* (The Hague: RPD, 1979). Review prepared for Dutch government of planning issues at European scale and role of international bodies in planning.
Ridley, F. F. (ed.) (1979), *Government and Administration in Western Europe* (Oxford: Martin Robertson).
Systems of public administration in Britain, France, Germany, Italy, Belgium and the Netherlands.
Thompson, J. M. (1978), *Great Cities and their Traffic* (London: Peregrine).

174 Planning in Europe

OECD study of thirty European cities.
Vanhove, N., and Klaasen, L. H. (1980), *Regional Policy, a European Approach* (Farnborough: Saxon House).
Whittick, H. (ed.) (1974), *Encyclopedia of Urban Planning* (New York: McGraw-Hill).
Includes entries on most European countries.
Wood, C., and Lee, N. (1978), *Physical Planning in the Member States of the European Economic Community*, Occasional Paper No. 2 (Manchester: Department of Town and Country Planning, University of Manchester).
Outline of planning statutes in the EEC of nine members.

(B) FEDERAL REPUBLIC OF GERMANY

Albers, G. (1974), 'Germany, West (Federal Republic)', in H. Whittick, *Encyclopaedia of Urban Planning* (New York: McGraw-Hill), pp. 459–83.
Bach, L. (1980), 'Urban and regional planning in West Germany', Exchange Bibliography No. 30, Council of Planning Libraries, Monticello, Illinois, USA.
A bibliography of source material in the English language.
Bowden, P. (1979), *North-Rhine Westphalia and North-West England: Regional Development in Action* (London: Anglo-German Foundation).
Comparative study of regional planning, and English translation of German regional programme.
Brösse, U. (1975), *Raumordnungspolitik* (Berlin: de Gruyter).
Childs, D., and Johnson, J.(1981), *West Germany – Politics and Society* (London: Croom Helm).
Crawley, A.(1973), *The Rise of Western Germany 1945–72* (London: Collins).
Dahrendorf, R. (1968), *Society and Democracy in Germany* (London: Weidenfeld & Nicolson).
Federal Ministry for Planning, Building and Urban Development (1975), *Regional Planning Programme for the General Spatial Development of the Federal Republic of Germany (Federal Regional Planning Programme)*, Regional Planning series No. 06.002 (Bonn and Bad Godesberg: Federal Ministry of Regional Planning, Building and Urban Development).
Federal Ministry for Planning, Building and Urban Development (1977), *Regional Planning and Urban Development: An Overview of the Federal Regional Planning Programme, the 1974 Regional Planning Report and the 1975 Urban Development Report* (Bonn and Bad Godesberg: Federal Ministry of Regional Planning, Building and Urban Development).
Hajdu, J. G. (1979), 'Phases in the post-war German urban experience', *Town Planning Review*, vol. 50, no. 3, pp. 267–86.
Hallett, G. (1976), *The Social Economy of West Germany* (London: Macmillan).
Introduction to Federal Republic of Germany.
Hallett, G. (1976), *Housing and Land Policies in West Germany and Britain* (London: Macmillan).
A comparative study, subtitled, 'A record of success and failure', with conclusions very favourable towards Germany.

Hass-Klau, C. (1982), *Planning Research in Germany* (London: SSRC).
Hellen, J. A. (1974), *North Rhine - Westphalia*, Problem Regions of Europe series (London: Oxford University Press).
Johnson, J., and Cochrane, A. (1981), *Economic Policy-Making by Local Authorities in Britain and Western Germany* (London: Allen & Unwin, 1981). Comparative study of selected local authorities, but more on Britain than Germany.
Kimminich, O. (1981), 'Public participation in the Federal Republic of Germany', *Town Planning Review*, vol. 52, no. 3 (July), pp. 274-9.
Krumme, G. (1974), 'Regional policies in West Germany', in N. M. Hansen (ed.), *Public Policy and Regional Economic Development. The Experience of Nine Western Countries* (Cambridge, Mass.: Ballinger), pp. 103-35.
Kunzmann, K. R., Estermann, H., and Rojahn, G. (1981), 'Development trends in the regional and settlement structure of Federal Germany', *Built Environment*, vol. 7, no. 3-4, pp. 243-54.
Kunzmann, K., and McLoughlin, J. B. (1981), 'Community and public authority responses to changing economic structure in older industrial areas', Ruhr Mersey Report 1, Dortmund and Liverpool.
Mellor, R. E. (1978), *The Two Germanies* (New York: Harper & Row). Geographical study of the two German states.
Michael, R., 'Metropolitan development concepts and planning policies in West Germany', *Town Planning Review*, vol. 50, no. 3, pp. 287-312.
Rosner, R. (1975), 'Town and regional planning in Germany', *Planner*, vol. 61, no. 10, pp. 375-8.
Solusbury, W. (1968), 'Local planning and the German Bebauungsplan', *Journal of the Town Planning Institute*, vol. 54, no. 3, pp. 117-22.
Waterhouse, A. (1979), 'The advent of localism in two planning cultures: Munich and Toronto', *Town Planning Review*, vol. 50, no. 3, pp. 313-25.
Williams, R. H. (1978a), 'Urban planning in Federal Germany', *Planner*, vol. 64, no. 2, pp. 46-7.
Williams, R. H. (1978b), 'Advocate planners', *Town and Country Planning*, December, pp. 553-7
Case study from Darmstadt, Federal Republic of Germany.

The German Federal Ministry responsible for planning publishes a series of planning research reports and accounts of planning policies, which are a valuable source of information, under the general title, *Schriftenreihe des Bundesministerium für Raumordnung, Bauwesen und Städtebau.*

(C) ITALY

Allen, K. J., and MacLennon, M. C. (1970), *Regional Problems and Policies in Italy and France* (London: Allen & Unwin).
Allen, K. J., and Stephenson, A. A. (1975), *An Introduction to the Italian Economy* (London: Martin Robertson).

Baldeschi, P. (1979), 'Participation and scientific method in Italian planning', *Planning Outlook*, vol. 20, no. 1, pp. 6–12.

Bianchi, G. (1979), 'Regional planning in Italy: a critical review', *Planning Outlook*, vol. 22, no. 1, pp. 2–5.

Cao-Pinna, V. (1974), 'Regional policy in Italy', in N. H. Hansen (ed.), *Public Policy and Regional Economic Development. The Experience of Nine Western Countries* (Cambridge, Mass.: Ballinger), pp. 137–79.

Fried, R. C. (1973), *Planning the Eternal City* (New Haven, Conn.: Yale University Press).
Politics and planning of Rome since the Second World War.

Haywood, J., and Watson, M. (eds) (1975), *Planning Politics and Public Policy: The British, French and Italian Experience* (Cambridge: Cambridge University Press).
Several contributors from Italy on economic, regional and other aspects of planning.

Scattoni, P., and Williams, R. H. (1978), 'Planning and regional devolution – the Italian case', *Planner*, vol. 64, no. 3 (March), pp. 38–40.

(D) FRANCE

Allen, K. J., and MacLennon, M. C. (1970), *Regional Problems and Policies in Italy and France* (London: Allen & Unwin).

Barrere, P., and Cassdou-Mounat, M. (1980), *Les Villes français* (Paris: Masson, 1980).

Bauchet, P. (1966), *La Planification française* (Paris: Editions du Seuil).

Burtenshaw, D. (1976), *Saar-Lorraine*, Problem Regions of Europe (London: Oxford University Press).

Chaline, C. (1981), 'Urbanization and urban policy in France', *Built Environment*, vol. 7, nos 3/4, pp. 232–42.

Flockton, C. H. (1982), 'Strategic planning in the Paris region and French urban policy', *Geoforum*, vol. 13, no. 3, pp. 193–208.

Hanley, D. L., Kerr, A. P., and Waites, N. H. (1979), *Contemporary France – Politics and Society since 1945* (London: Routledge & Kegan Paul).

Hayward, J., and Watson, M. (eds) (1975), *Planning, Politics and Public Policy: The British, French and Italian Experience* (Cambridge: Cambridge University Press).
Several contributors from France on economic, regional and other aspects of planning.

House, J. W. (1978), *France – an Applied Geography* (London: Methuen).

Kain, R. (1982), 'Europe's model and exemplar still: the French approach to urban conservation 1962–82', *Town Planning Review*, vol. 53, no. 4 (October), pp. 403–22.

Liggins, D. (1975), *National Economic Planning in France* (Farnborough: Saxon House).

Lucas, N. J. D. (1979), *Energy in France: Planning Politics and Policy* (London: Europa).

Macrory, R., and Lafontaine, M. (1982), *Public Inquiry and Enquête Publique* (London: ENDS). Comparative study of English and French public inquiry practice.
Meny, F. (1982), 'Urban planning in France', in D. H. McKay (ed.), *Planning and Politics in Western Europe* (London, Macmillan), pp. 13-41.
Merlin, P. (1976), *Les Villes nouvelles françaises* (Paris: La Documentation Française).
Prud'homme, R. (1974), 'Regional economic policy in France', in N. H. Hansen (ed.), *Public Policy and Regional Economic Development. The Experience of Nine Western Countries* (Cambridge, Mass.: Ballinger), pp. 33-63.
Rubenstein, J. M. (1978), *The French New Towns* (Baltimore, Md: Johns Hopkins University Press).
Sutcliffe, A. (1970), *The Autumn of Central Paris: The Defeat of Town Planning 1850-1970* (London: Edward Arnold).
Thompson, I. B. (1973), *The Paris Basin*, Problem Regions of Europe series (London: Oxford University Press).
Thompson, I. B. (1975), *The Lower Rhône and Marseilles*, Problem Regions of Europe series (London: Oxford University Press).
Tuppen, J. (1979), 'New towns in the Paris region - an appraisal', *Town Planning Review*, vol. 51, no. 1 (January), pp. 55-70.

In addition, issues of current interest in French planning are regularly discussed in the quarterly journal *Etudes foncières*.

(E) THE NETHERLANDS

Amsterdam City Council (1975), *Amsterdam Planning and Development* (Amsterdam: Public Works Department, Town Planning Section).
Bigham, A. (1973), 'Town and country planning in Britain and the Netherlands', *Journal of Planning and Environmental Law*, pp. 294-301.
Brussard, W. (1979), *The Rules of Physical Planning* (The Hague: Ministry of Housing and Physical Planning).
Constandse, A. K. (1976), *Planning and Creation of an Environment; Experiences in the Ijsselmeerpolders* (Lelystad: National Agency for the Ijsselmeerpolders).
Ester, P. (1981), 'Environmental concern in the Netherlands', in T. O'Riordan and R. K. Turner (eds), *Progress in Resource Planning and Environmental Management* (London: Wiley), pp. 81-108.
Faludi, A. (1978), *Promotion and Control of Development in the Netherlands* (Amsterdam: Institute for Planning and Demography of the University of Amsterdam).
Faludi, A., and Hamnett, S. (1979a), *The Promotion and Control of Development in Leiden*, Working Paper No. 27 (Oxford: Department of Town Planning, Oxford Polytechnic, 1979).
Faludi, A., and Hamnett, S. (1979b), *Flexibility in Dutch Local Planning*, Working Paper No. 28 (Oxford: Department of Town Planning, Oxford Polytechnic).

Hamnett, S. (1975), 'Dutch planning – a reappraisal', *Planner*, vol. 61, no. 3 (March), pp. 102–5.
Hamnett, S. (1978), 'Leiden-Nerenwijk: a case study of Dutch local planning', *Planning and Administration*, vol. 5, no. 2, pp. 28–42.
Hamnett, S. (1982), 'The Netherlands: planning and the politics of accommodation', in D. H. McKay, *Planning and Politics in Western Europe* (London: Macmillan), pp. 111–43.
Heide, H. ter (1979) *Implications of Current Demographic Trends for Population Redistribution Policies* (The Hague: Ministry of Housing and Physical Planning).
Hendriks, A. J. (1974), 'Regional policy in the Netherlands', in N. M. Hansen, *Public Policy and Regional Economic Development. The Experience of Nine Western Countries* (Cambridge, Mass.: Ballinger), pp. 181–97.
Lawrence, G. R. P. (1973), *Randstad Holland*, Problem Regions of Europe (London: Oxford University Press).
McLintock, H., and Fox, M. (1971), 'The Bijlmermeer development and the expansion of Amsterdam', *Journal of the Town Planning Institute*, vol. 57, pp. 313–16.
Minett, J. (1978), 'Local planning in the Netherlands', *Planner*, vol. 64, no. 2 (March), pp. 41–3.
Minett, J. (1979), *Local Planning in Netherlands and England*, Working Paper No. 37 (Oxford: Department of Town Planning, Oxford Polytechnic).
Ministry of Housing and Physical Planning (1960), *First Report on Physical Planning* (The Hague: Ministry of Housing).
Ministry of Housing and Physical Planning (1966), *Second Report on Physical Planning* (The Hague: Ministry of Housing).
Ministry of Housing and Physical Planning (1976), *Summary of the Report on Urbanisation* (The Hague: Ministry of Housing).
Ministry of Housing and Physical Planning (1977), *The Human Settlement Policy in the Netherlands* (The Hague: Ministry of Housing).
Ministry of Housing and Physical Planning (1978a), *Summary of Rural Areas Report* (The Hague: Ministry of Housing).
Ministry of Housing and Physical Planning (1978b), *Transport Policy and the Development of the Transportation Network in the Netherlands* (The Hague: Ministry of Housing).
Ministry of Housing and Physical Planning (1980), *Housing in the Netherlands: Facts and Figures* (The Hague: Ministry of Housing).
Needham, B. (1982), *Choosing the Right Policy Instruments* (Epping: Gower Press).
Local authority employment policies in three Dutch *Gemeenten*, and in the UK.
Pinder, D. (1976), *The Netherlands* (London: Hutchinson).
Postma-van Dijck, J. E. J. M. (1978), *Monitoring and Adjustment of Regional Plans* (Delft: Planning Research Centre/TNO).
Steigenga, W. (1973), *Ranstad Holland: Concept in Evolution (Planologie in beweging)* (Amsterdam: Institute for Planning and Demography of the University of Amsterdam).
Wrathall, J. E., and Carrick, R. J. (1980), 'Flexible planning in the Netherlands: new town development in the Ijsselmeerpolders', *Planning Outlook*, vol. 23, no. 3, pp. 114–21.

Whysall, P., and Beynon, N. J. (1980) 'Redevelopment in Amsterdam', *Planning Outlook*, vol. 23, no. 2, pp. 77-82.

In addition, material on Dutch planning (in English)) may be found in the following periodicals: *ECE Newsletter*, edited by the National Physical Planning Agency; *Planning and Development in the Netherlands*, edited by the Institute of Social Studies, The Hague.

(F) BELGIUM

Albrechts, L. (1981), 'Organisational structure versus emergence of plans in Belgian planning', *Planning Outlook*, vol. 24, no. 2, pp. 41-8.
Brussels Agglomeration Authority (1979), *Urbanisme à Bruxelles, Petit Guide 1979* (Brussels: Administration de l'urbanisme).
Riley, R. (1976), *Belgium* (London: Hutchinson).
 An economic geography.
Royaume de Belgique (1979), *Loi du 29 Mars 1962 de l'aménagement du territoire et de l'urbanisme* (Brussels: Co-ordination officiense, October).
 Basic town planning law of 1962 and amendments up to 1979.
Suetens, L. P. (1981), 'Public participation in Belgium', *Town Planning Review*, vol. 52, no. 3 (July), pp. 267-73.
Tips, W., and Sallah, W. (1979), 'Belgium structure planning', *Town and Country Planning*, vol. 48, no. 3, pp. 90-1.

(G) LUXEMBOURG

Edwards, K. C. (1968), *Luxembourg: The Survival of a Small Nation* (Nottingham: Department of Geography, University of Nottingham).
Luxembourg (1974), *Journal Officiel du Grand Duché*, A-No. 18 (23 March), pp. 309-15.
 Contains the text of the National Planning Act.
Poos, J. F. (1977), *Crise économique et les petites nations – le modèle luxembourgeois* (Luxembourg: Centre des recherches européenes).

In general, for reading on Luxembourg, it is necessary to rely on the appropriate chapters of books listed in section A such as Clout (1981) or Wood and Lee (1978), or in section L such as Yuill, *et al.* (1980) or Yuill and Allen (1982).

(H) UNITED KINGDOM

Alexander, A. (1982), *Local Government in Britain since Reorganisation* (London: Allen & Unwin).
Ashworth, W. (1954), *The Genesis of Modern British Town Planning* (London: Routledge & Kegan Paul).
Blowers, A. (1980), *The Limits of Power* (Oxford: Pergamon).
 Analysis of planning in Bedfordshire.

Cherry, G. E. (ed.) (1974a), *Urban Planning Problems* (London: Leonard Hill, 1974).
Cherry, G. E. (ed.) (1974b), *The Evolution of British Town Planning* (London: Leonard Hill, 1974).
Cherry, G. E. (ed.) (1982), *The Politics of Town Planning* (London: Longman).
Cross, D. R., and Bristow, M. R. (eds) (1982), *English Structure Planning* (London: Methuen).
Cullingworth, J. B. (1973), *Problems of an Urban Society*, 3 vols (London: Allen & Unwin).
Cullingworth, J. B. (1979), *Essays on Housing Policy* (London: Allen & Unwin).
Cullingworth, J. B. (1980), *Environmental Planning 1939–69*, 4 vols (London: HMSO).
Cullingworth, J. B. (1982), *Town and Country Planning in Britain*, 8th edn (London: Allen & Unwin).
Eversley, D. (1973), *The Planner in Society* (London: Faber).
Fagence, M. (1977), *Citizen Participation in Planning* (Oxford: Pergamon).
Gibson, M. S., and Langstaff, M. J., (1982), *An Introduction to Urban Renewal* (London: Hutchinson).
Glasson, J. (1981), *An Introduction to Regional Planning*, 2nd edn (London: Hutchinson).
Goldsmith, M. (1981), *Politics, Planning and the City* (London: Hutchinson).
Hall, P., Gracey, H., Drewett, R., and Thomas, R. (1973), *The Containment of Urban England*, 2 vols (London: Allen & Unwin).
Hall, P. G. (1982), *Urban and Regional Planning*, 2nd edn (Harmondsworth: Penguin).
Healey, P. (1979), *Statutory Local Plans*, Working Paper No. 36 (Oxford: Department of Town Planning, Oxford Polytechnic).
Hebbert, M. (1980), 'The British new towns – a review article', *Town Planning Review*, vol. 51, no. 4 (October), pp. 414–20.
HMSO (1969), *People and Plans* (London: HMSO).
The Skeffington Report on public participation.
HMSO (1979), *Environmental Planning in Britain*, COI Reference Pamphlet No. 9 (London: HMSO).
Howard, E. (1902, 1965), *Garden Cities of Tomorrow* (London: Faber).
Johnson, J., and Cochrane, A. (1981), *Economic Policy-Making by Local Authorities in Britain and Western Germany* (London: Allen & Unwin).
Lawless, P. (1979), *Urban Deprivation and Government Initiative* (London: Faber).
Lawless, P. (1981), *Britain's Inner Cities: The Problem and its Solution* (London: Harper & Row).
McAuslan, P. (1975), *Land Law and Planning* (London: Weidenfeld & Nicolson).
McAuslan, P. (1980), *The Ideologies of Planning Law* (Oxford: Pergamon).
McKay, D. H. (1982), 'Regulative planning in the centralised British state', in D. H. McKay (ed.), *Planning and Politics in Western Europe* (London: Macmillan).
McKay, D. H., and Cox, A. W. (1979), *The Politics of Urban Change* (London: Croom Helm).
McLouglin, J. B. (1973), *Control and Urban Planning* (London: Faber).

Osborn, F. J., and Whittick, A. (1977), *New Towns, their Origins, Achievements and Progress* (London: Routledge & Kegan Paul).
Ratcliffe, J. (1982), *An Introduction to Town and Country Planning*, 2nd edn (London: Hutchinson).
Richards, P. G. (1980), *The Reformed Local Government System*, 4th edn (London: Allen & Unwin).
Roberts, J. F. (1976), *General Improvement Areas* (Farnborough: Saxon House).
Self, P. (1982), *Planning the Urban Region* (London: Allen & Unwin).
Spence, N., Gillespie, A., Goddard, J., Kennett, S., Pinch, S., and Williams, H. (1982), *British Cities – an Analysis of Urban Change* (Oxford: Pergamon).
Williams, R. H. (1982), 'Tyne and Wear's industrial improvement areas', *Northern Economic Review*, vol. 1, no. 4 (September), pp. 18–27.

(I) IRELAND

Buchanan, C. D. (1969), *Regional Development in Ireland* (Dublin: An Foras Forbatha).
Roberts, P. W. (1975), *Regional Planning in the Republic of Ireland* (Liverpool: Department of Town and Country Planning, Liverpool Polytechnic).

In addition, the annual reports of the Planning Board, An Bord Pleanala, are a useful source of reference.

(J) DENMARK

Byplan Labratorium (1982), *Town Planning Guide* (Copenhagen: BL).
Skovsgaard, K.-J. (1982), 'Danish planning and consensus politics', in D. H. McKay, (ed.). *Planning and Politics in Western Europe* (London: Macmillan), pp. 144–69.
Travis, A. S., and Hamblin, J. (1978), *Denmark: A Case Study in Tourism, Development and Environmental Conservation*, Research Memorandum No. 69 (Birmingham: Centre for Urban and Regional Studies, University of Birmingham).

(K) GREECE

Amourgis, S. (1979), 'Planning and regional development policy in Greece: reality or myth?', *Epikentra*, nos. 9–10, pp. 57–65; in Greek.
Aravantinos, A. (1970), 'Planning objectives in modern Greece', *Town Planning Review*, vol. 41, no. 1, pp. 41–62.
ESYE (1977), *Statistical Yearbook of Greece*, National Statistical Service of Greece (Athens: ESYE); in Greek and English.
KEPE (1976a), *Development Program 1976-80. Vol. 10, Regional Development* (Athens: KEPE); in Greek.

KEPE (1976b), *Development Program 1976-80. Vol. II, Urban Organisation* (Athens: KEPE); in Greek.
Kontogiorgis, G. (ed.) (1977), *Social and Political Forces in Greece* (Athens: Political Science Society of Greece/Exantas); in Greek.
Kyriakou-Skassi, A. (ed.) (1975), *Housing in Greece. Realisations in the Private Sector* (Athens: Ministry to the Prime Minister/TEE); in Greek.
Lagopoulos, A.-Ph. (1977), *Offices*, Vol. 7 of *Research on Urban Standards* (Athens: Centre for Urban Planning Research, National Technical University); in Greek.
Lagopoulos, A. -Ph. (1978), 'The relation between social formation and settlement network in Greece, 1821-1971' (also to appear in English), in E. Andrikopoulou-Kafkala, V. Dimitriadis, G. Kafkalas, P. Papadopoulou, and A.-Ph. Lagopoulos (co-ordinators), *Study of the Urban Settlement Network in Greece*, Phase 3 (Thessaloniki: TEE/Chair B of Urban Planning, University of Thessaloniki); in Greek.
Leontidou-Gerardi, K. (1979), 'Problems of the organisation of the network of urban centres in the country', *Synchrona Themata*, vol. 2, no. 6, pp. 17-32; in Greek.
Loukakis, P. (1979), *Programme of Economic and Social Development 1978-1982. Preliminaries* (Athens: National Printing Office); in Greek.
TEE (1975), *Housing in Greece: Government Activity* (Athens: TEE); in Greek, English, French and Russian.
Wassenhoven, L. (1970), 'Development poles and the policy of the competitive cities', *Oikonomia ke Koinonia*, vol. 1, no. 4, pp. 32-9; in Greek.

(L) THE EUROPEAN COMMUNITIES

This section contains Commission publications of particular significance, accounts of the European Communities institutions and procedures ranging from introductory to detailed descriptive and analytical works, and reviews of specific areas of EEC policy-making. Not many books on the European Communities contain much reference to town and country planning, but these works will enable planning to be set in the wider context of EC policy-making and the concept of European integration..

Arbuthnot, H., and Edwards, G. (eds) (1979), *A Common Man's Guide to the Common Market* (London: Macmillan).
Camag, R., and Cappellin, R. (1981), 'European regional growth and policy issues for the 1980s', *Built Environment*, vol. 7, no. 3-4, pp. 162-71.
Commission of the European Communities (1977), *First Report on the State of the Environment* (Brussels: CEC).
Commission of the European Communities (1979a), *Second Report on the State of the Environment* (Brussels: CEC).
Commission of the European Communities (1979b), *A Transport Network for Europe*, Bulletin Supplement 8/79 (Brussels: CEC).
Commission of the European Communities (1980a), *Progress Made in Connection with the Environment Action Programme*, COM (80) 222 Final Brussels: CEC).

Commission of the European Communities (1980b), *Draft Directive concerning the Assessment of the Environmental Effects of Certain Public and Private Projects*, COM (80) 313 Final (Brussels: CEC, June).
Commission of the European Communities (1981), *Draft Action Programme on the Environment 1982-86*, COM (81) 626 Final (Brussels: CEC, November).
Commission of the European Communities (1982), *Proposal to Amend the Proposal for a Council Directive concerning the Assessment of the Environmental Effects of Certain Public and Private Projects*, COM (82) 158 Final (Brussels: CEC, March).
El-Agraa, A. M. (ed.) (1980), *The Economics of the European Community* (Oxford: Philip Allan).
Gwilliam, K. M. (1980), 'The transport policy', in A. M. El-Agraa (ed.), *The Economics of the European Community* (Oxford: Philip Allan), ch. 8.
Henig, S. (ed.) (1979), *Political Parties in the European Community* (London: Allen & Unwin).
Henig, S. (1980, *Power and Decision in Europe* (London: Europotentials Press). Structure and role of institutions in practice.
HMSO (1981), *Report of the House of Lords Select Committee on the European Communities: Environmental Assessment of Projects. 11th Report 1980-81*, House of Lords Paper No. 69 (London: HMSO). Text of the proposal. All the evidence and a very thorough report.
Hudson, R., and Rhind, D. (1983), *An Atlas of EEC Affairs* (London: Methuen).
Hull, C., and Rhodes, R. A. W. (1977), *Inter-Governmental Relations in the European Community* (Farnborough: Saxon House).
Includes discussion of public participation in Germany and the UK.
Kerr, A. J. C. (1977), *The Common Market and How It Works* (Oxford: Pergamon).
Klein, L. (1981), 'The European Community's regional policy', *Built Environment*, vol. 7, no. 3-4, pp. 182-9.
Kormoss, I. B. F. (1974), *The European Community in Maps* (Brussels: CEC).
Lasok, D., and Bridge, J. W. (1976), *An Introduction to the Law and Institutions of the European Communities*, 2nd edn (London: Butterworths).
Lee, N., and Wood, C. (1978), 'EIA – a European perspective', *Built Environment*, vol. 4, no. 2, pp. 101-10.
Lodge, J. (ed.) (1983), *Institutions and Policies of the European Community* (London: Frances Pinder).
Nevin, E. T, (1980), 'Regional policy', in A. M. El-Agraa (ed.), *The Economics of the European Community* (Oxford: Philip Allan), ch. 13.
Noel, E. (1979), *The European Community. How it Works* (Brussels: CEC).
Pinder, D. (1983), *Regional Economic Development and Policy. Theory and Practice in the European Community* (London: Allen & Unwin).
Roberts, P.W. (1981), 'Regional policy – the European perspective', *Town Planning Review*, vol. 52, no. 3 (July), pp. 325-33.
Swann, D. (1978), *The Economics of the Common Market* (Harmondsworth: Penguin).
Tsoukalis, L. (1981), *The European Community and its Mediterranean Enlargement* (London: Allen & Unwin).

Context for Greece, Spain and Portugal.
Twitchett, C. C. (1981), *Harmonisation in the EEC* (London: Macmillan).
Wallace, H., Wallace, W., and Webb, C. (1977); *Policy Making in the European Communities* (London: Wiley).
Detailed analysis of political process by which policies are formulated. Includes regional, but not environment, policy.
Whittaker, M. (1979), 'The relationship between local government and the EEC', paper presented to Regional Studies Association Conference on Finance for Regional Development, Dublin, 26–27 June 1979.
Wohlfahrt, J. (1979), 'The European Economic Community – expectations and realities of integration', in D. J. Haywood and R. N. Berki (eds), *State and Society in Contemporary Europe* (Oxford: Martin Robertson), ch. 10.
Yuill, D., and Allen, K. (eds) (1982), *European Regional Incentives 1982* (Strathclyde: Centre for Study of Public Policy, University of Strathclyde).
Includes information on each member-state (10), plus Portugal, Spain and Sweden, and an overview.
Yuill, D., Allen, K. and Hull, C. (eds) (1980), *Regional Policy in the European Community* (London: Croom Helm).
Chapters on each member of the EEC of nine.

(M) THEORY AND COMPARATIVE METHODOLOGY

Breakell, M. J. (ed.) (1975), *Problems of Comparative Planning*, Working Paper No. 21 (Oxford: Department of Town Planning, Oxford Polytechnic).
Papers presented at a conference in Oxford.
Davies, H. W. E. (1980), *International Transfer and the Inner City*, Occasional Paper No. 5 (Reading: School of Planning Studies, University of Reading).
Summary of findings and transfer methodology employed in trinational study (UK, USA and the Federal Republic of Germany).
Faludi, A., and Hamnett, S. (1975), *The Study of Comparative Planning*, CES Conference Paper No. 13 (London: CES).
Review of cross-national studies and other theoretical papers.
Hamnett, S. (ed.) (1974), *The Nature and Purpose of Comparative Planning* (Delft: Delft University of Technology).
Masser, I. (1981a), 'The analysis of planning processes: a framework for comparative research', Paper TRP 28, Department of Town and Regional Planning, University of Sheffield, January.
Masser, I. (1981b), 'Comparative planning studies. A critical review', Paper TRP 33, Department of Town and Regional Planning, University of Sheffield, October.
Perroux, F. (1964), *La Notion de pôle de croissance* (Paris: Presses Universitaires de France).
Exposition of the growth-pole concept which has underlain policy in a number of countries.
Sharpe, L. J. (1975), 'Comparing planning policy: some cautionary comments', in M. J. Breakell, *Problems of Comparative Planning*, Working Paper No. 21 (Oxford: Department of Town Planning, Oxford Polytechnic), pp. 26–32.
White, P. M. (1978), 'Towards an improved methodology for cross national comparative planning research', CURS Working Paper No. 62, University of Birmingham.

Index

Index

Abercrombie 11, 49, 162
Action area plans 90
Advisory committees of Experts 71, 79, 136, 169
Almere 54
Amsterdam 49, 54
Antwerp 65
Appeals 92, 108, 110
Archaeological sites 77, 136
Architecture profession 19, 20, 46, 47, 59, 61, 66, 69, 126, 140, 141, 157, 168
Athens 129-31, 134, 135-6, 142

BDA 19
Bebauungsplan 18, 21, 22, 163
Belgium 63-72, 84, 144, 147, 160, 161-3, 166, 167
Berlin 5
Bestemmingsplan 51, 56-8, 163
Birmingham 88, 98
Bonn 11
Bouw plan 58
Brandt 13
Brittany 39
Brussels 67, 145, 146, 167

Charter of Athens 15
Christaller 11, 66
CIAM 49
Civil engineering profession 44, 46, 48, 61, 126, 157, 168
Cologne 73
Commission of the European Communities 145-9, 154-5, 156, 157-8, 164, 171
 Directorates - general 146, 149
Comparative planning 4-6, 159, 169
Conservation 41, 95, 112-13, 136, 165-6
Copenhagen 115-16, 124
Cork 105, 106
Council of Europe 170
Council of Ministers 145, 147-8, 153
County councils 80, 105, 108-9, 120, 123

County manager 106, 108, 109

DATAR 38, 44
DDE 38, 42-4, 45
Delta plan 54
Denmark 1, 114-27, 145, 160, 162, 164, 165, 167
Department of the Environment 89, 103
Development control 6, 43, 70, 91-2, 106, 107-9, 111, 164
Development corporation 40, 93, 165
Devolution 26, 31, 32, 64, 69, 70, 118, 167-8
District plans 90, 163
Doxiadis 141
Dublin 105, 113
Dunkirk 38

Economic and Social Committee 145, 147-8, 152-3, 169
Economic planning 9, 22-4, 31-2, 38, 66, 90, 117, 121-2, 140-1, 169-70
 Economic planning councils 89
Employment planning 99
England 88-9, 90-1, 103
Enterprise zone 100-1, 164
Entwicklungs plan 22-4
Environment fund 151
Environmental-impact analysis 41, 43, 149, 152-5, 164, 170
Environmental protection 24, 53, 54, 56, 80, 133, 150, 152, 170
Ettlingen 13
Euratom 2, 144, 145
European Administrative Centre, Kirchberg 78-9
European Coal and Steel Community (ECSC) 2, 144-5, 149, 155-6
European Council (EC) 147
European Economic Community (EEC) 1-3, 37, 77, 84, 129, 133, 142-3, 144-60, 162
European Investment Bank 77, 155-6
European Parliament 145, 147-8, 152

188 Index

European Regional Development Fund 133, 146, 149, 154, 162
European Social Fund 155
Examination-in-public 89

Flächennutzung plan 18, 21-2, 163
Flanders 66-7, 71, 167
Flemish Federation of Planners 69, 169
Fordergebiete 11
Fos 39
France 37-48, 75, 138, 141, 144, 146, 153, 154, 161, 162, 163, 164, 166-7, 168, 169
Frankfurt 12, 165

Geddes 49
General improvement areas 96-7, 99
Gent 67-8, 69, 71
Germany, Federal Republic 5, 8-25, 74-5, 84, 138, 141, 144, 145, 160-1, 162-3, 164-5, 167, 170
  Basic Law 10, 16
  constitution 16, 167
Giunta 32
Glasgow 93
Greece 1, 3, 128-43, 145, 154, 162, 165, 166, 169
Green belt 94
Green Party (ecological) 10, 41, 161
Gronigen 52, 53
Growth pole 13, 23, 38, 50, 66, 135, 142, 162

Hague, The 53, 54
Hamburg 12
Hanover 12
Hessen 19
Historic buildings 41, 92, 136, 165
Horsfall 170
Housing policies 22, 52, 64, 95, 109, 130, 135, 137, 140, 165, 170
  Housing action areas 97

Industry policies 66, 74, 80, 82, 100, 130-1, 136, 140-1, 152, 155
  Industrial improvement areas 99-100
Inner-city policies 93, 98-9, 135
Institute of Urban Planners 32, 33-4

Ireland 1, 103-13, 145, 154, 163, 164, 165, 167
Irish Planning Institute 113, 168
Island Authorities 88, 91
Italy 3, 26-36, 144, 146, 154, 160, 164, 167, 168
  constitution 28, 36
  regions 28-9, 32

KEPE 133, 134
Køge 116, 165

Landscape plans 23
Leeds 88
Lelystad 54
Limburg 53
Limerick 105, 106
Liverpool 93, 98, 99, 157
Local plans 90-1, 125-6, 163
London 73, 88, 93, 98-9, 160, 162
Luxembourg City 73-85, 144, 162, 164, 166, 169
Luxembourg Compromise 148

Manchester 88, 99
Marshall Fund 9, 10
Master-plan 138, 140, 143
Métropole d'équilibre 39
Mitterrand 48, 166, 167
Monnet 144
Munich 12

National parks 39, 44, 46, 47, 86, 87, 94
National Physical Planning Agency 55-6, 160
National planning 37, 55, 57, 63, 123, 140, 162
Netherlands, The 49-62, 84, 144, 160, 162, 163
  Crown 58
  constitution 56
Newcastle-upon-Tyne 88, 93, 99, 102
New towns 162, 165
  Belgian 65
  British 40, 87, 93, 94
  Danish 116
  French 40, 44
  German 11

Index 189

NIROV 61
Nord-West Stadt 165
Northern Ireland 88, 89, 154
Northrhine-Westphalia 8-9, 16, 18-19

OP (planned development area) 142
OREAM 38, 44

Paris 38, 39, 40, 45
Paris Summit 147, 154
Parliament
  Belgian 70
  British 87-8
  Danish 115, 117, 123
  Dutch 58
  French 38, 48
  Greek 132-3
  Irish 103
  Luxembourg 76, 77
PASOK 128, 132, 139, 141
Pedestrianisation 5, 12, 95, 97, 166, 170
Perroux 66, 142
Planning appeals 43, 47
Planning board 107-11, 163, 167
Planning consultants 17, 45, 55, 63, 69, 77, 87, 126, 135, 166
Planning education 20, 46, 48, 61-2, 66, 69, 102, 113, 126, 141-2, 168
Polders 52-4, 55
Portugal 145
POS 39-40, 42, 45, 46, 47, 163
Préfet 38, 40, 42-4, 47
PRGC 27-9, 30-1, 32, 33, 34
PRI 34
Process planning 60
Professional planners 45-6, 47-8, 60, 61-2, 69, 87, 101-2, 113, 135, 157-8, 162, 166, 168
PTC 28, 34
Public inquiries 43, 68
Public participation 13, 19, 25, 32, 47, 71, 101, 107, 119, 121, 127, 140, 161, 166

Randstad 50, 52, 53
Rechtstaat 51, 163

Regional planning 11, 26, 33-4, 53, 57, 63, 66-7, 116, 118-19, 120, 123-5, 134, 138, 140-1, 143, 154, 157, 162
Rijkswaterstaat 56
Rotterdam 54, 166
Royal Town Planning Institute 87, 101-2, 113, 157, 168-9
Ruhr 12, 15, 24
Rural planning 23, 52-3, 54, 56, 64, 79, 80, 94, 112-13, 117, 122-3, 136

Schumann 144
Scotland 88-9, 91, 93
SDAU 39, 41-2, 46
SFU 46
Sheffield 88
Social housing 10, 22, 64
Social sciences 20, 45, 60, 141
Sozial plan 22
Spaak 144
Spain 145
SRL 19
Strasbourg 147
Strategic planning 162
Streekplan 57, 162
Structure plans 57, 67, 69, 89-91, 116-18, 125, 162, 167
Structuurplan 57, 163
Subject plans 90

Terminology 6-7, 8, 169
Thessaloniki 129-30, 134, 135, 136, 142
Tourism 39, 44, 80, 137, 141, 152
Transport planning 22, 24, 45, 56, 64, 65, 75, 79, 116, 124-5, 131, 156, 170
Treaty of Paris 146
Treaty of Rome 144, 146-7, 148-50, 156-7
Tübingen 13

Ulm 13
United Kingdom 1, 3, 6, 86-102, 141, 145, 147, 153-5, 161, 162-4, 165, 167
Universities 20, 46, 63, 66, 67, 69, 102, 113, 135, 141

Unwin 49
Urban Development Corporation 93, 100
Urban renewal 18, 22, 32, 41, 52, 54, 58, 61, 65, 67, 95-6, 114-15
USA 10-12, 129, 130
Utrecht 53, 54

Volkswagen Works 11

Wales 88-9, 90, 103
Wallonia 66, 167
War 9, 12, 26, 29, 33, 70, 75, 76, 78, 129, 144, 160

ZAC 40, 46
ZAD 40, 46
ZEP 138-9

For Product Safety Concerns and Information please contact our EU
representative GPSR@taylorandfrancis.com
Taylor & Francis Verlag GmbH, Kaufingerstraße 24, 80331 München, Germany

www.ingramcontent.com/pod-product-compliance
Lightning Source LLC
Chambersburg PA
CBHW052118300426
44116CB00010B/1712